Advances in Gravity Concentration

Advances in Gravity Concentration

Editor

Gyan Singh

scitus
academics

Advances in Gravity Concentration

Edited by **Gyan Singh**

Printed in 2017

ISBN: 978-1-68117-472-3

Library of Congress Control Number: 2015936590

© 2016 by
SCITUS Academics LLC,
616, Corporate Way, Suite 2, 4766,
Valley Cottage, NY 10989

www.scitusacademics.com

Contents

vi

Preface

Gravity has been an essential force in mineral and coal processing for centuries. While some newer separation technologies developed during the industrial age are widely used, gravity-based separators remain the prominent means of producing concentrates from coal, iron ore, rare earths, industrial minerals, tin, and tungsten ores. Recent advances in gravity-based separation technologies have reduced previous particle size limitations and improved their effectiveness in treating mixed-phase particles as compared to competing technologies. It is now possible to achieve efficient, high-capacity gravity separations for ultrafine particles using gravity-based units that provide centrifugal forces many times that of natural gravity. Sophisticated research equipment provides opportunities for in-situ studies of processes, resulting in new fundamental theories and control schemes.

Editor

Heavy Metals' Spatial Distribution Characteristics in a Copper Mining Area of Zhejiang Province

Hua Sun[1], Juan Li[1], and Xiaojun Mao[2]

[1]Department of Resource, Environment and Urban and Rural Planning, Nanjing Agricultural University, Nanjing, China

[2]Department of Geographic Science, Zhejiang Normal University, Jinhua, China

ABSTRACT

The spatial distribution characteristics of six heavy metals and metalloid in soil of Zhuji Lipu copper mining area, Zhejiang Province, was studied by using geostatistics approaches combined with GIS. These elements included Pb, As, Cr, Cu, Zn and Ni. The statistical analyses showed that concentrations of these elements were lognormal distribution. Concentrations of Pb, As, Cu, Zn and Ni were strongly correlated with each other indicating that these elements in soils may be from the same pollution source. However, accumulation of Cr was unique with its geometric mean being close to that in the control soil. This indicates that Cr content was mainly influenced by soil factors. The Kriging method was applied to estimate the unobserved points. The Kriging interpolation maps reflected significant spatial distribution of these elements as influenced by both pollution and geological factors. The present study indicated that GIS based geostatistics method could accurately analyze the spatial variation of heavy metals and metalloid in the mining area. Overall, higher concentrations of heavy metals and metalloid were found in the center of both the north and south sides. The content of copper in the south was significantly higher than that in the north due to paddy field land uses. In addition, the terrain of four terraces tilted to the center and the broad irrigation accident occurred in the 4th trench in the south of sampling area were also contributed to the higher concentrations of these elements.

INTRODUCTION

Arable land pollution was becoming more serious in China and the world [1]. Mining and smeltering activities not only generated heavy metal pollution but also resulted in determination of soil quality and destruction of the ecological function of the system [2]. Currently in southern region of China, rapid development in mining industry has produced significant environmental problems including mine soil, water, and air pollution [3-6], which seriously affected food security and health of human and animal. Due to their long-term duration, heavy metal polluted soil is difficult to be remediated [7]. Therefore, remediation of heavy metal polluted soils has become a hot issue and understanding of the spatial distribution and variation of heavy metal pollution is an important prerequisite.

The objectives of this study were to investigate the spatial distribution and variation of heavy metals and metalloid in soils of Hang Pu copper plants of Zhuji City, Zhejiang Province of China.

STUDY AREA AND RESEARCH PROGRAMS

The Study Area

Zhejiang North-central Pu mine is located in Zhuji City, Zhejiang Province, 29°43›23»N, 119°59›09»E with a total area of the mining area 0.8 km 2. Hilly terrain within the mining area was a major landscape. Mining land is in the subtropical monsoon climate with annual average temperature of 16.2°C and ≥10°C annual accumulated temperature 4924°C - 5233°C. Annual rainfall is 1335.9 mm and annual average evaporation is 1260.7 mm with annual average relative humidity of 75.1%. Soil type is mainly yellow red soil with red sandstone-shale development. Sampling areas are located north of the mining area.

Soil Sampling Plan

Since the mine area had relatively simple topography, the total 48 samples with a space between adjacent points of 10 meter were planned in order to ensure a representative sampling. In the north side of one major irrigation trench, all these sampling were carefully distributed according to the land-use types and topographic characteristics (face, direction and length of ladder status) (Figure 1).

Mining water overflowing has occurred in the south east side of the 4th irrigation tunnels. Sampling areas was located about 300 meters south side and irrigation canal was along the general direction of flowing through the sampling areas from south to north central region.

Soil Sample Collection and Analyses

According to "soil environmental quality monitoring technical specifications (NY/T395-2000)", soils were collected and prepared. Surface soil (0 cm - 20 cm) was sampled from each point. Soils were then air-dried and ground through 2 mm sieve for analyses. Soils were analyzed for total lead, arsenic, chromium, copper, zinc, nickel, pH and etc [8].

Lead, arsenic, chromium, copper, zinc, nickel were determined with British X-ray fluorescence wavelength dispersive spectrometer (Model AFS-820, PANalytical Axios production company).

Spatial Variability Analysis Method

Geographic information system (GIS) is fit for acquisition and processing of the environmental information and based on geographic location and the spatial analysis techniques, which can make a quantitative evaluation of regional environment. We can visual display changes in the regional distribution of pollution by taking advantage of GIS technology with visualization of the results of the evaluation. At present, geographic information system (GIS) has been widely applied to the research of regional environmental pollution as an important spatial analysis technology; it provides effective means for researching the spatial variability of pollutants.

Therefore, the paper chooses this method to research the spatial distribution characteristics of heavy metals and show the importance of the impact of heavy metals on the ecological environment.

In this paper, geostatistical spatial analyses were used to generally semi-variance map and describe spatial variability [9]. As a best spatial interpolation method, it has been widely applied to a regionalized variable characterization in soil science, environmental science and ecology and other fields [10]. ArcGIS9.1 was used for this geostatistical analysis, including the semi-variogram (semivariogram) of the calculation and fitting comparison, Kriging spatial interpolation and simulation error analysis.

Semi-variance function is also often referred to as variation function, which is studied in geostatistics of soil variability [11]. Located in the one-dimensional (two-dimensional or three-dimensional) space in different locations x_1, x_2, $\cdots$ x_n on a certain soil characteristics of the observed value of $Z(x_1)$, $Z(x_2)$, $\cdots$ $Z(x_n)$, Semi-variance [(h)] can reflect the spatial dependence of regionalized variables, the calculation can be estimated under:

$$\gamma(h) = \frac{1}{2N(h)} \sum_{i=1}^{n} \left(Z(X_i) - Z(X_i + h) \right)^2$$

where N (h) is used for spacing the number of all pairs of observation points. With h as the abscissa, (h) was for the vertical coordinate mapping, that is, semi-variance diagram. Spatial local interpolation (Kriging method) is based on variation of a function and its structural analysis in a limited region of the regionalized variable values for the best linear unbiased estimation method, namely, (is a weighting factor, Z(x) is the sample value, Z is the estimated value):

$$z = \sum_{i=1}^{n} \lambda_i z\left(x_i\right)_i$$

RESULTS AND ANALYSIS

Statistical Analysis of Concentrations of Soil Heavy Metals and Metalloid

Concentrations of six heavy metals were summarized in Table 1. The results showed that the highest variation was found for concentrations of As, Cu, Zn, and Pb. The sources of these elements were contamination from mining activities over the years, especially through acid mine waste water. According to field surveys in 1967, the study area received acid mine waste water outside the Bay. In addition, a heavy application of chemical fertilizers and pesticides may also be important pollution sources for these elements. The major chemical fertilizer was superphosphate, potassium chloride, and urea. Frequently phosphate fertilizers contained some of these elements. Among all these elements, Cr had the least variation as well as Ni, indicating fewer impacts of anthropogenic sources on these two elements Cr and Ni.

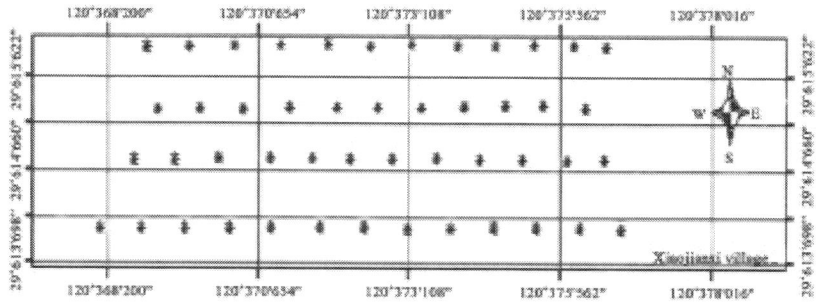

Figure 1: Distribution of sampling points.

Table 1: Descriptive statistics of soil heavy metals and metalloid contents (mg/kg).

Item	Max	Min	Median	Average	Standard deviation	Skewness	Kurtosis	Distribution Type
As	37.9820	1.3820	64.3070	70.5432	39.6218	-0.04078	1.7586	Lognormal
Cr	116.1390	53.3990	97.4500	95.2812	12.5296	-1.16260	4.8448	Lognormal
Cu	1095.9320	90.9040	702.8885	664.5950	249.3063	-0.45336	2.2543	Lognormal
Zn	9212.4770	369.5700	5542.7550	4691.7190	2237.9530	-0.29452	2.0588	Lognormal
Ni	61.8510	18.8130	47.9535	45.7578	11.4556	-0.65865	2.4018	Lognormal
Pb	2076.7610	152.8280	1024.9320	1072.2480	565.3570	-0.11522	-1.6825	Lognormal

According to the soil environmental quality standard of China (GB15618-1995), soils in the studied area (pH < 6.5) exceeded 100% of copper and zinc, lead standards, 91.67% of arsenic and 79.17% of nickel. For the studied area, soil copper, zinc, lead and arsenic contents have reached critical levels of pollution. But nickel pollution was relatively slight with an average of 45.7 mg/kg (Ni Class II standard is 40 mg/kg), while chromium has not yet to reach pollution levels. Therefore, the study area was in general characterized by combined pollution of main Cu, Zn, Pb and As.

In addition, the skewness and kurtosis analyses indicate that arsenic, chromium, copper, zinc, nickel, and lead had a negative bias, and the addition to both Cr spikes outside the state. The frequency distribution histogram (Histogram) showed that six heavy metals and metalloid in soils studied were characterized by normal distribution.

The Correlation between Heavy Metals and Metalloid in Soils

Table 2 showed that in soils of the studied area, a strong cross correlation occurred among these heavy metals and metalloid, especially between As-Cu, As-Zn, As-Ni, AsPb, Cu-Zn, Cu-Ni, Cu-Pb, Zn-Ni, Zn-Pb, Ni-Pb. This indicates that all these elements may have a same source mainly from anthropogenic mining relevant activities. However, correlation between Cr and As or Cr, Cu, Zn, Ni, and Pb were low. Therefore, As has the homology with Cu, Zn, Ni and Pb.

Affected by mining pollution processes, the content of As changed simultaneously with other heavy metals, while Cr was not significant, the elements of mining pollution and the distribution were further demonstrated.

Spatial Variation Analysis of Heavy Metal in the Soil under the Conditions of the Isotropic

Due to the presence of specific value of the variable, which may cause interruption of a continuous surface, experimental semi-variogram will distort or even obscure the inherent spatial structure of the variable characteristics [12]. In this paper we identified the specific valuedomain method [13], that is, the sample average plus or minus three times the standard deviations, in this interval (±3 s) other than the data were as specific values, and then the normal maximum and minimum values were used instead of specific values. Subsequent calculations were based on the field data after treatment.

A study of six soil heavy metals and metalloid in the isotropic condition was shown in Figure 2. According to the actual variation of the function, semi-variance for the vertical axis and the sampling distance of abscissa displayed the semi-variance fitting curve drawn map. Each element of the semi-variogram curve has a significant continuous increase in range. When the semi-variance with the increasing of separation distance has reached a certain scale (variable range), the Semi-variance curve becomes flat. The correlation between samples in more than this distance was no longer relevant after, and it appeared as when h > a time, (h) was at a value of up and down swing, so there is spatial variability of the structure. This change pattern can be used with a threshold value (Sill) models fitted to choose the best theoretical model variogram. Using ArcInfo in the analysis of the statistical analysis module, a variety of model interpolation error may be produced. If the forecast errors are unbiased, then the mean prediction error should be close to zero.

Table 2: The correlation coefficients of concentrations of soil heavy metals and metalloid

	As	Cr	Cu	Zn	Ni	Pb
As	1	0.451	0.910**	0.817**	0.875**	0.971**
Cr	0.451**	1	0.445	0.402	0.505	0.501
Cu	0.910**	0.455	1	0.832**	0.903**	0.873**
Zn	0.817**	0.402	0.882**	1	0.846**	0.779**
Ni	0.875**	0.505	0.903**	0.846**	1	0.870**
Pb	0.971 **	0.501	0.873**	0.779**	0.870**	1

**indicates at p < 0.01 (n 48).

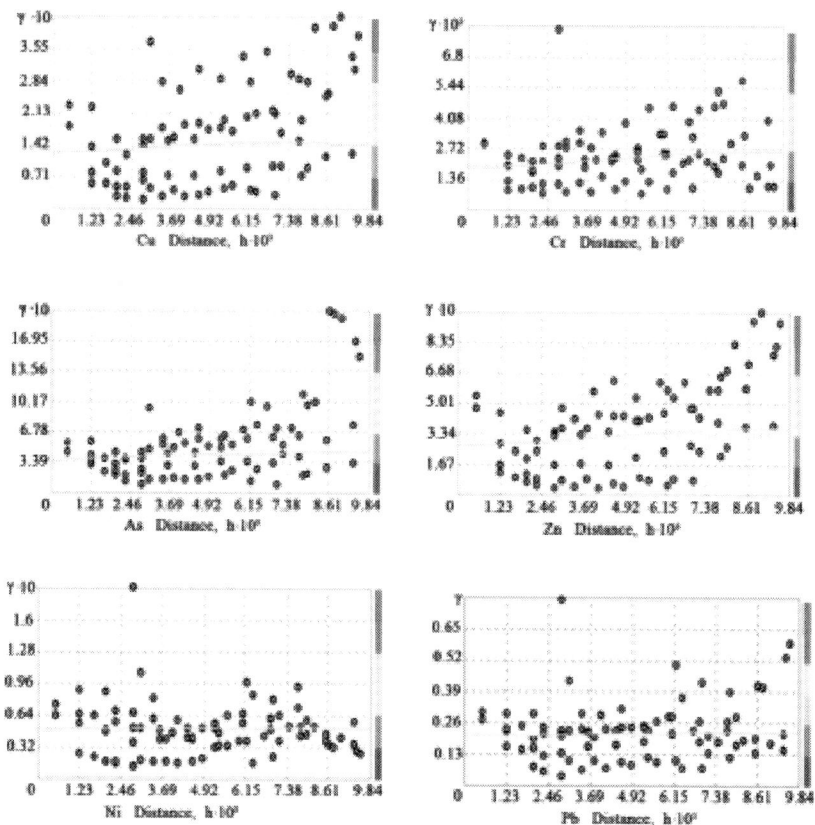

Figure 2: The corresponding graph of semivariograms of soil heavy metals.

It is necessary to fit different methods and parameters repeatedly for determining the theoretical model, in order to obtain the theoretical variation function for simulating that distance has impacted on data point. So, theoretical variation curves should be the most approximation with experimental variation of the function, which was generated by the chosen method, especially with small steps. It is required to determine the fitting degree between the theoretical variation function and the experimental variation function for inspecting the validity of the model.

Through several times of fitting, Table 3 shows the best fitting theoretical model and its parameters for the six heavy metals and metalloid in the semi-variance selected. The results show that, for Cr and Zn the Gaussian model may be theoretical model of the semi-variance; the J-Bessel may be for Ni; the Hole Effect model for Cu and Pb; and the Exponential model for As.

The ratios of soil nugget values to the value of base stations for these elements were as follows: Lead > Copper > Arsenic > Zinc > Nickel > Chromium. Lead, copper, and arsenic had ratios of element nugget value to that of base stations, of 92.6014%, 86.6664%, 76.0163%, respectively. The ratios of soil nugget values to those of base stations for other three heavy metals were 25% - 75%. Spacial correlations were controlled by the structural factors and random factors. Structural factors included such as parent material, soil type, climate, and soil-forming factors, while random factors were farming, management practices, cropping systems, land use patterns, pollution and other human activities.

The optimized model should meet the following criteria: the minimized (close to 0) Standard Mean (Mean Standardized), the minimum root mean square prediction error (Root-Mean-Square), the minimized (close to 0) average standard error (Average Standardized Errors), closest root mean square prediction Error (Root-Mean-Square), closest (to 1) standard root mean square prediction error (Root-Mean-Square-Standardized) [14] (Table 4).

The Simulated Values and Measured Values of the Cross-Validation

Cross-validation was conducted in the data of all samples. Each time removing one of the points, with the remaining points, the value of the predicted value of the point, was compared to the actual value and the predictive value of the difference between the validations of spaceinterpolation was used to analyze the degree of accuracy, thus the best Kriging analysis was selected. In theory, the best prediction is equal to the actual measured values and the slope between them should be a linear related. As the spatial interpolation in the process of smoothing (smoothing) effects, the linear slope between the measured values and predicted values is usually less than 1. For example, the predicted values of soil Pb and the measured values of the relevant equation are:

Y = 0.71X + 342.779 The correlation between measured values and the predictive values of these six heavy metals and metalloid was at significant level (Table 5).

Table 3: Theoretical semivariogram models of soil heavy metals and metalloid and their corresponding parameters

Element	Theoretical model	Variable-range	Sill	Nugget	Nugget/sill
Cr	Gaussian	9626.5	2.70824	1.8997	0.701452
Ni	J-Bessel	1956.2	0.68449	0.48857	0.713772
Cu	Hole effect	9626.5	1.45235	1.2587	0.866664
Zn	Gaussian	9626.5	3.7619	2.7505	0.731146
As	Exponential	9626.5	4.6073	3.5023	0.760163
Pb	Hole effect	7681.8	0.22207	0.20564	0.926014

Table 4: The interpolation errors of semivariogram corresponding models

Element	Root-Mean-Square	Average-Standardized Errors	Mean Standardized	Root-Mean-Square Standardized
Cr	12.23	13.93	—0.01525	0.8814
Ni	8.995	11.27	0.01924	0.8861
Cu	3 204.3	278.5	—0.02274	0.9998
Zn	2117	3472	—0.1064	1.284
As	36.83	71.32	—0.03516	0.8842
Pb	411.1	632.2	—0.02677	1.003

Table 5: The correlation coefficients between measured values and predictive values.

Soil Properties	As	Cr	Cu	Zn	Ni	Pb
Correlation	0.655**	0.368**	0.617**	0.511**	0.626**	0.728**

**At $p = 0.01$ level.

The Spatial Distribution of Soil Heavy Metals and Their Patterns

According to the principles of Kriging interpolation and semi-variogram fitting parameters, Geostatistical Analyst in ArcMap software was applied for analyzing module spatial variability of interpolation. Therefore spatial distribution of trends of six heavy metals and metalloid and the distribution of these elements classification map were produced (Figure 3). Kriging interpolation was affected by variation function model simulation accuracy, the distribution of samples, the selection of the number of adjacent samples and many other factors. Interpolation accuracy may be reflected through the interpolation error. The smaller the variance, i.e., the standard root mean square prediction error close to 1, the more accurate Kriging interpolation [15].

The Kriging interpolation maps for soil copper, zinclead, arsenic, and nickel content were shown in Figures 3-8. The overall distribution trends show a north-south direction with higher values in the center. With decreasing from the center of north-south direction the higher values were found in the south than in the north of Grand.

According to the Chinese 1995 soil environmental quality standards, the maximum allowable concentration of pollutants were summarized as follows (Figures 3-8). In all regions copper and zinc contents were more than Class II standard criteria for acidic soils; in northwest, east and south-east part of the region, the lead level was between Class I and Class II threshold criteria, and the rest of area, Pb was beyond the critical value of the standard Class II for acidic soil. In addition, the north-west and eastern parts of the region had arsenic content around the threshold criteria of Class I and the rest areas had As beyond the

Class II criteria. Northwest, northeast, east and south-east areas had nickel content between the standard critical values of Class I and II for acidic soil with the rest areas exceeding threshold of Class I and Class II.

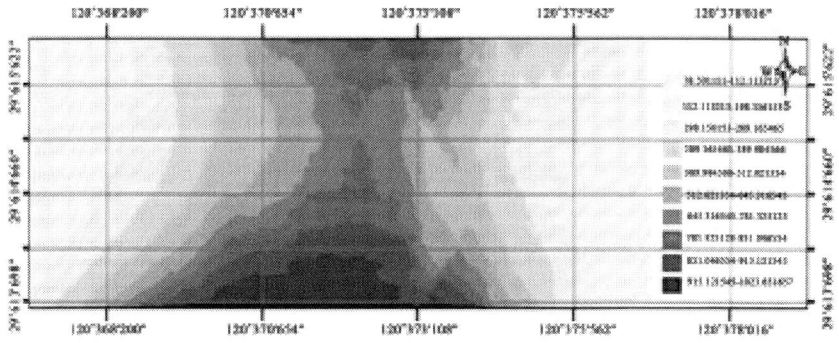

Figure 3: The interpolation map of Cu.

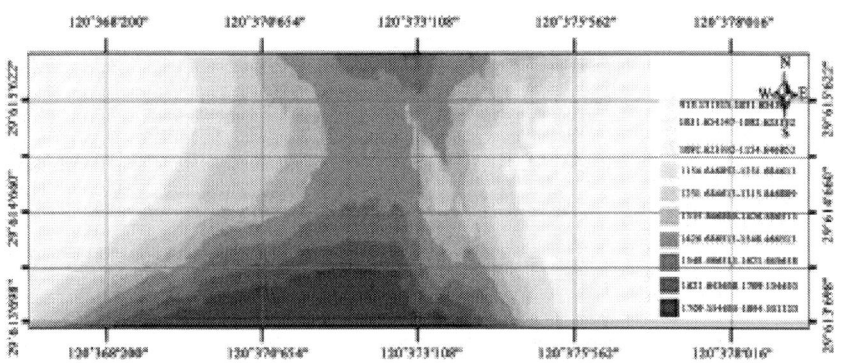

Figure 4: The interpolation map of Zn.

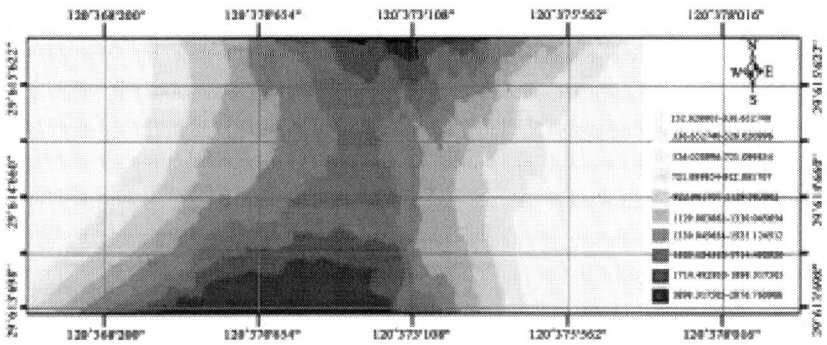

Figure 5: The interpolation map of Pb.

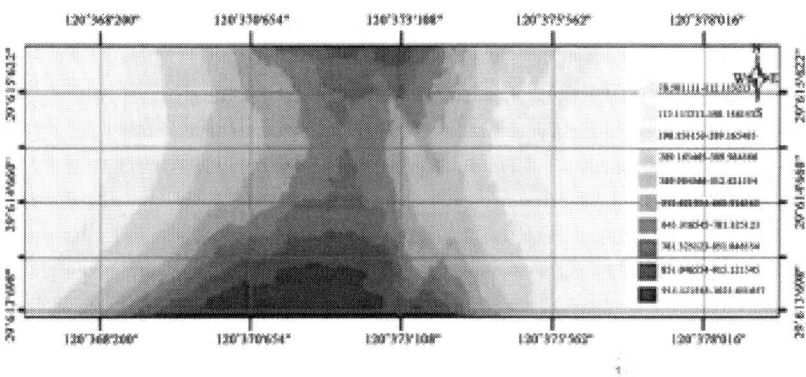

Figure 6: The interpolation map of As.

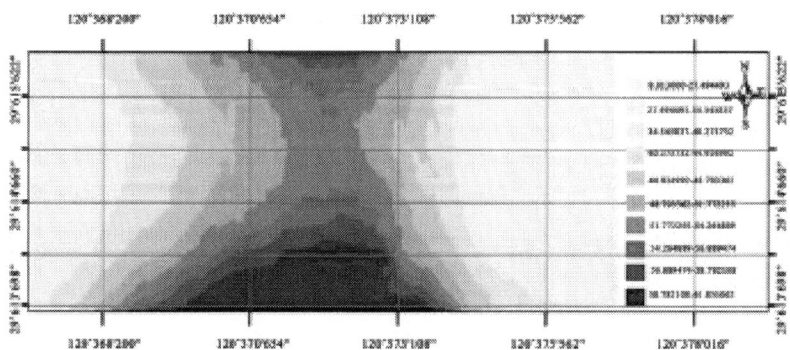

Figure 7: The interpolation map of Ni.

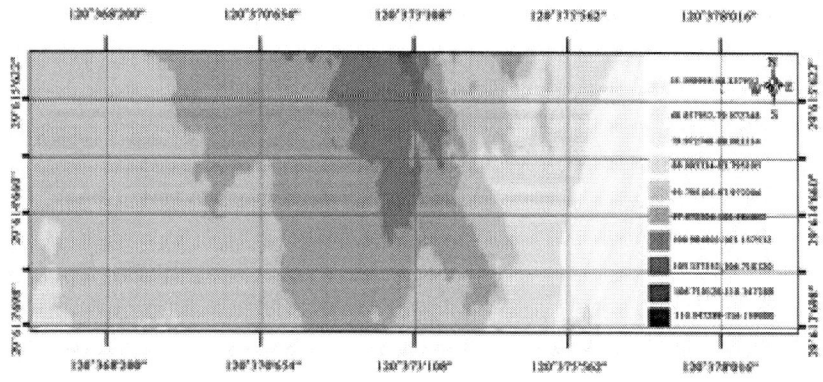

Figure 8: The interpolation map of Cr.

The overall distribution of chromium in soil showed the trends with highest values in the north, followed by the south and decreasing in the direction of the east-west. In the study area, north and south appeared to have relatively high values of chromium. East, south-east and north-east regions with mainly mountainous areas had relatively low chromium content. According to the 1995 soil environmental quality standards, the east, southeast and northeast regions had chromium contents below the Class I critical value, while the rest of regions had Cr contents between Class I and Class II criteria for the acidic soil. All soils had Cr concentrations below the standard threshold for acidic soil.

CONCLUSIONS

Heavy metals and metalloid (arsenic, copper, zinc, nickel, and lead) have been significantly accumulated in soils around the 4th trench. Strong correlation among these elements may indicate all these elements came from the same pollution sources [16]. The contamination sources may be from both natural, i.e., geochemical sources, or anthropogenic source, i.e., human activities, or combined pollution sources. The present study showed strong correlation between As-Cu, As-Zn, As-Ni, As-Pb, Cu-Zn, Cu-Ni, Cu-Pb, Zn-Ni, Zn-Pb, Ni-Pb, indicating the highly possibility of the same sources of these elements. However, correlation between Cr and As, or Cr, Cu, Zn, Ni, and Pb was low, indicating that Cr was less accumulate in the region as well as less possibility of exogenous anthrogenic source. On the other hand, As, Cu, Zn, Ni, Pb were the elements most associated with pollution source.

Ratios of nugget values of heavy metals in soils to base stations were in the order: lead > copper > arsenic > zinc > nickel > chromium. Lead, copper, arsenic had the ratios of nugget values to base stations of 92.6%, 86.6%, and 76.0%, respectively. In accordance with classification criteria of spatial correlation of regionalized variables [17], these three elements were specially weakly correlated, because human activity weakened their spatial correlation. The rest heavy metals had 25% - 75% ratios of nugget values to base stations, indicating their moderate spatial correlation. Chromium had relatively low ratio of nugget values to base stations and the chromium content in soils of studied area was close to the geometric mean, implying that chromium was mainly affected by soil formation factors.

Copper, zinc, lead, arsenic, and nickel had the similar spatial distribution in the studied areas. In general, their distribution showed the high accumulation in the center of the regions with decreasing concentrations form the center towards both south and north direction. In the study area, higher concentrations of Cu were found in the center region in the direction of south-north, and concentration of Cu in the south side was higher than that in the north side. This is mainly due to land use patterns and topography of the landscape in the region. Paddy fields were main land use in the south side where irrigation water was through the 4th trench. However, in the north side, vegetable and wasteland were the major land use where runoff flew from the mining tailing area. This may contribute to the high concentrations in the soils of the north side. In addition, the terrain of four terraces tilted to the center and the broad irrigation accident occurred in the 4th trench in the south of sampling area were also contributed to the higher concentration of these elements.

REFERENCES

1. J. M. Shu, J. J. Wang and X. C. Liu, "Ecological Restoration of Mining Wasted Lands," China Population, Resource and Environment, Vol. 8, No. 3, 1998, pp. 72-75.

2.   M. K. Zhang and Z. X. Ke, "Copper and Zinc Enrichment in Different Size Fraction of Organic Matter from Polluted Soils," Pedosphere, Vol. 14, No. 1, 2004, pp. 27-36.

3.   H. M. Chen, "Heavy Metals Pollution in Soil-Plant System," Science Press, Beijing, 1996, pp. 1-14.

4.   Y. J. He and L. X. Zhu, "China's Mineral Resources," Shanghai Education Press, Shanghai, 1987, pp. 219-220.

5.   L. Gao, W. M. Zhang and X. H. Yang, "Studies on the Issues of Modern Ecology: The Rectification of Nonferrous Metals Industrial Environment and the Reclamation of Mine Land," China Science and Technology Press, Beijing, 1996, pp. 572-582.

6.   C. Tu, C. R. Zheng and H. M. Chen, "The Current Status of Soil-Plant System in Copper Mine Tailings," Acta Pedologica Sinica, Vol. 37, No. 2, 2000, pp. 135-143.

7.   Z. M. Gao, "Ecological Research on the Pollution of SoilPlant System," China Science and Technology Press, Beijing, 1986.

8.   Institute of Soil Science, "Analysis of Soil Physical and Chemical Properties," Shanghai Scientific and Technical Publishers, Shanghai, 1978.

9.   T. M. Burgess and R. Webster, "Optimal Interpolation and Isarithmic Mapping of Soil Properties. I. The SemiVariogram and Punctual Kriging," Journal of Soil Science, Vol. 31, No. 2, 1980, pp. 315-331. doi:10.1111/j.1365-2389.1980.tb02084.x

10.   P. Goovaerts "Geostatistics in Soil Science: State-of-theArt and Perspectives," Geodema, Vol. 89, No. 1-2, 1999, pp. 1-45. doi:10.1016/S0016-7061(98)00078-0

11.   H. C. Guo, "Land Restoration in China," Acta Ecologica Sinica, Vol. 10, No. 1, 1990, pp. 24-26.

12.   C. S. Zhang, S. Zhang and J. B. He, "Spatial Distribution Characteristics of Heavy Metals in the Sediments of Yangtze River System-Geostatistics Method," Acta Geographica Sinica, Vol. 52, No. 2, 1997, pp.185-192.

13.   S. Q. Wang, S. L. Zhu and C. H. Zhou, "Characteristics of Spatial Variability of Soil Thickness in China," Geographical Research, Vol. 20, No. 2, 2001, pp. 161-169.

14. H. B. Li, Z. H. Lin and S. X. Liu, "Application of Kriging Technique in Estimating Soil Moisture in China," Geographical Research, Vol. 20, No. 4, 2001, pp. 446-452.

15. G. A. Tang and X. Yang, "ArcGIS Spatial Analysis Experiment Guide," Science Press, Beijing, 2006, p. 405.

16. F. A. Galley and O. L. Lioyd, "Grass and Surface Soils as Monitors of Atmospheric Metal Pollution in Central Scotland," Water, Air and Soil Pollution, Vol. 24, 1985, pp. 1-18.

17. C. A. Cambardella, T. B. Moorman, J. M. Novak, et al., "Field-Scale Variability of Soil Properties in Central Iowa Soils," Soil Science Society of America Journal, Vol. 58, No. 5, 1994, pp. 1501-1511. doi:10.2136/sssaj1994.03615995005800050033x.

Chapter 2

Volcanic Ash versus Mineral Dust: Atmospheric Processing and Environmental and Climate Impacts

Baerbel Langmann

Institute of Geophysics, University of Hamburg, Geomatikum, Office 1411, Bundesstraße 55, 20146 Hamburg, Germany

ABSTRACT

This review paper contrasts volcanic ash and mineral dust regarding their chemical and physical properties, sources, atmospheric load, deposition processes, atmospheric processing, and environmental and climate effects. Although there are substantial differences in the history of mineral dust and volcanic ash particles before they are released into the atmosphere, a number of similarities exist in atmospheric processing at ambient temperatures and environmental and climate impacts. By providing an overview on the differences and similarities between volcanic ash and mineral dust processes and effects, this review paper aims to appeal for future joint research strategies to extend our current knowledge through close cooperation between mineral dust and volcanic ash researchers.

INTRODUCTION

Volcanic ash represents a major product of volcanic eruptions [1–3]. It is formed by fragmentation processes of the magma and the surrounding rock material of volcanic vents [1, 4]. Depending on the strength of a volcanic eruption, volcanic ash is released into the free troposphere or even the stratosphere [1, 5], where it is transported by the prevailing winds until it is removed from the atmosphere by gravitational settling and wet deposition [6]. Volcanic ash is also known to be mobilised by wind from its deposits [7–12], which have accumulated after volcanic eruptions on land located along the main transport directions of the volcanic cloud, which spreads out over hundreds to thousands of kilometres, dependent on wind speed, ash size, ash density, and eruption magnitude. In contrast to atmospheric mineral dust, the importance of volcanic ash for climate has long been considered negligible [5].

The global mineral dust cycle and its interactions with the Earth's climate system have been studied widely [13–18]. Mineral dust aerosols affect the radiative forcing of the atmosphere directly [13, 19] and indirectly by acting as cloud condensation or ice nuclei [20, 21]. Furthermore, mineral dust aerosols influence ozone photochemistry [22, 23] and supply nutrients to marine [24] and terrestrial ecosystems [25]. Vice versa, climate variability affects the mineral dust burden of the atmosphere through modifications of precipitation, vegetation cover, and wind [15].

This review contrasts the environmental and climatic effects of volcanic ash versus those of mineral dust. A stronger focus is put on the description of volcanic ash, whereas mineral dust effects are described in less detail, but with referencing the extensive literature. Similarities and differences will be emphasised (Figure 1) to facilitate the different scientific communities studying volcanic ash and mineral dust to learn from each other in an interdisciplinary way, to think about future joint research projects, and to address the important, challenging, and compelling questions, which are still open such as the following:

- which physical-chemical processes during long-range transport in the atmosphere affect the surface chemical composition of volcanic ash and mineral dust?

- how important is resuspension of volcanic ash from deposits on land for posteruptive climate and environmental effects?

- what are the reasons for the huge variability of nutrient and toxic element fluxes from volcanic ashes and mineral dust to the surface ocean?

- how important is volcanic iron fertilisation of the surface ocean and the associated modifications of atmospheric CO_2 in comparison to that induced by mineral dust?

- how relevant is the role of volcanic ash and/or mineral dust for the Earth's climate?

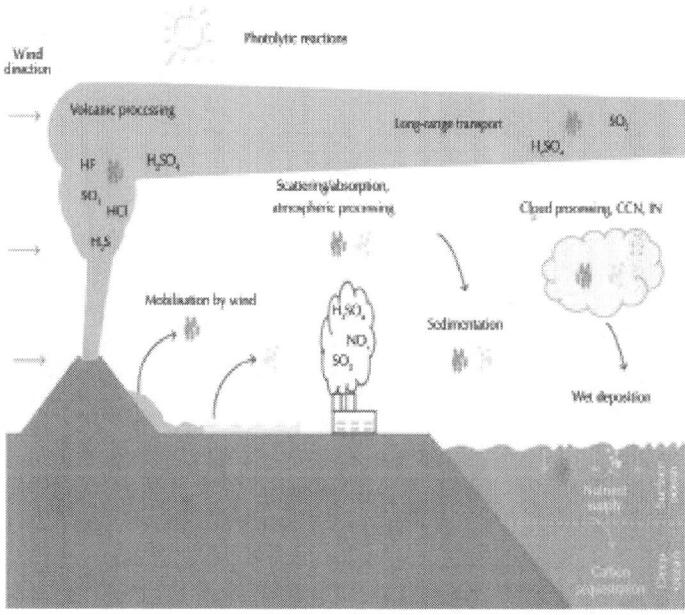

Figure 1: Schematic diagram showing the important processes controlling environmental and climate effects of volcanic ash (in grey) and mineral dust (in yellow). (CCN: cloud condensation nuclei; IN: ice nuclei).

Section 2 gives definitions for mineral dust and volcanic ash and provides information of the general chemical and physical properties. Sources, atmospheric load, and deposition processes are discussed in Section 3, atmospheric processing in Section 4, and environmental and climatic impacts in Section 5. The last section summarises needs for future research.

DEFINITIONS AND CHEMICAL AND PHYSICAL PROPERTIES

General Definition

According to the "Glossary of Atmospheric Chemistry Terms" [26], dust consists of small, dry, and solid particles released into the atmosphere by natural forces, such as wind, volcanic eruptions, and by mechanical or man-made activities (e.g., crushing, milling, and shoveling). Dust particles are usually in the size range from about 1 to 100 μm in diameter and settle slowly from the atmosphere due to gravity [26]. Thus, mineral dust and volcanic ash may constitute a fraction of all recorded dust. To distinguish between mineral dust and volcanic ash the following definitions are used: atmospheric mineral dust originates from a suspension of minerals constituting the soil, whereas volcanic ash is loose and unconsolidated material with particle diameters less than 2 mm [27] being either dispersed in the atmosphere or being deposited above the soil. Weathered mineral dust may originate from volcanic tephra (tephra is defined as any fragmental material produced by a volcanic eruption regardless of composition and fragment size; [28]); however, volcanic ash represents relatively fresh material produced during a recent volcanic eruption (recent in this context means no longer than about 100 years ago) and is therefore different form mineral dust.

Chemical Composition

The main chemical elements contained in mineral dust, as well as volcanic ash, are silicium and oxygen, which constitute the main components of minerals and rocks in the Earth's crust and mantle.

The chemical composition of the bulk volcanic ash is mainly determined by the magma from which it is generated. Generally, three types of magma are distinguished from each other (Table 1). These types of magma have different melting points, viscosities, and typical volatile contents. The mineral composition of volcanic ash consists of about 45–75 wt% of silica [27]. In addition, silicate is the main component of most minerals like feldspar, olivine, pyroxene, hornblende, and biotite [29]. These minerals are formed through successive crystallisation during cooling and decompression when the magma ascends from the Earth's mantle through the Earth's crust into the conduit and subsequently into the volcanic plume [28]. During the crystallisation process, the composition of the melt is changing due to depletion of crystallised components and enrichment of remaining components driving the successive generation of different minerals, including those without silicate, such as magnetite or ilmenite.

Table 1: Major types of magma

Magma type	SiO_2 (wt%)	T_{melt} (°C)	Viscosity and gas content
Basaltic	45–55	1000–1200	Low
Andesitic	55–65	800–1000	Intermediate
Rhyolitic	65–75	650–1000	High

A major difference in the mineral composition of mineral dust and volcanic ash results from chemical weathering of mineral dust, generally on geological time scales. Mineral composition changes under the influence of water, oxygen, and acids. For example, feldspar will weather to clay and iron-bearing minerals can form hematite and goethite. Therefore primary iron containing minerals (e.g., amphiboles and pyroxenes) of volcanic rocks are not identified in mineral dust [30]. Further multiphase chemical modifications at the surfaces of mineral dust and volcanic ash take place during atmospheric transport (see Section 4).

Physical Properties

Size Distribution

Close to the source regions, the size distribution of mineral dust particles varies from about 0.1 to over 100 µm in diameter, depending on soil characteristics and wind speed [18]. The particle size spectrum of emitted mineral dust largely controls the fraction that is transported over large distances. The coarser the particles, the faster they are deposited. However, to characterise long-range transported mineral dust, precise particle size measurements for less than about 10 µm are still a challenge [31].

The size distribution of volcanic ash is greatly dependent on the formation process, which is either an explosive volcanic eruption, a phreatomagmatic eruption, or a pyroclastic density current. In addition, secondary volcanic ash clouds result from the resuspension from volcanic ash depositions on land (see Section 3.1), which should not be mixed up with coignimbrite clouds as discussed later. Explosive volcanic eruptions occur when magma containing dissolved volatiles rises in the conduit. Thereby exsolution of volatiles forms gas bubbles that grow by diffusion, decompression, and coalescence. The further the magma-gas mixture rises, the more the pressure decreases, leading to an acceleration of the mixture against gravitational and friction forces, until a continuous gas stream with clots and clasts of magma (called pyroclasts) leaves the vent explosively [1]. The explosive character of a volcanic eruption depends considerably on the viscosity of the magma. In general, most efficient fragmentation occurs during explosive eruptions where magmas of rhyolitic composition are involved because of the higher volatile content (Table 1).

Phreatomagmatic eruptions [1] are triggered by the interaction of external water with magma, for example, from a glacier as during the early phase of the Eyjafjallajökull eruption [32]. External water may also be supplied from crater lakes or even the shallow ocean during seamount volcanic eruptions [33]. Highly efficient fragmentation is caused by thermal contraction of magma due to chilling on contact with water [4]. Water initially chills the magma at the interface, which then shatters. The water penetrates the mass of shattered hot glass and is transformed into high-pressure superheated steam by a runaway process of heat transfer and further magma fragmentation, until a violent explosion results. Violent phreatomagmatic eruptions produce especially fine-grained volcanic ash.

Pyroclastic flows occur when the eruption column or lava dome collapses leading to gas and tephra flows rushing down the flanks of a volcano at high speed, which thereby also contribute to the fragmentation process through milling by the collisional processes [1, 34]. Coignimbrite clouds can arise from pyroclastic flows when the material at the top of a pyroclastic flow gets more buoyant than the surrounding air. These convective clouds can form volcanic plumes as high as the original feeding plume and are a source of substantial amounts of fine volcanic ash as well.

As eruption conditions may be highly variable in time, all fragmentation processes can take place simultaneously. According to [2] volcanic ash particles with diameters smaller than 1 mm contribute about 55–97 wt% to the total ash content. Volcanic ash particles with diameters less than 30 μm make up only a few weight percent during basaltic eruptions, whereas they can contribute 30–50 wt% to the total ash content during rhyolitic eruptions. However, it is the even finer particle size fraction (PM_{10} and $PM_{2.5}$) that may be carried for hundreds of kilometres before settling onto land or into the ocean.

Density and Surface Area

The morphology of mineral dust particles can be assumed to be spherical with a widely used particle density of 2650 kg/m^3 in mineral dust modelling, for example, [35]. The density of individual volcanic ash particles varies from one eruption to another and even during an eruption. Generally it lies between 2000 and 3000 kg/m^3 [28,36] dependent on the basaltic or rhyolitic composition, the amount of crystallisation, and porosity [37]. Due to the expansion of magmatic gases like H_2O, CO_2, SO_2, H_2, CO, H_2S, HCl, and HF [38] during an explosive and phreatomatic volcanic eruption (see Section 2.3.1), volcanic ash generally consists of vesicular particles with an undifferentiated surface texture. Generally, the specific surface area of volcanic ash is smaller than 2 m^2/g [39]. However, values up to 10 m^2/g have also been reported [39].

EMISSIONS INTO THE ATMOSPHERE, ATMOSPHERIC LOAD, AND SUBSEQUENT DEPOSITION

Emissions

The notation "emission" describes the release of material from outside the atmosphere into the atmosphere, where the location outside the atmosphere represents a source for the atmosphere. Mineral dust source areas are generally located in semiarid or arid areas where the surface is sparsely vegetated and dry. Here, fine grained material can accumulate and be mobilised into the atmosphere by wind. Numerical models for mineral dust mobilisation usually define dust emission areas based on, for example, soil moisture [40], soil texture [13], and vegetation effects [13, 41]. Mineral dust emissions into the atmosphere are a complex, nonlinear function of both soil surface properties (size distribution of the surface soils, roughness length of erodible and nonerodible particles, and soil moisture) and meteorological conditions (wind friction velocity and precipitation). Mineral dust emissions from an erodible surface occur when the wind friction velocity exceeds a threshold value, dependent on the soil properties [42]. However, by using different assumptions, for example, erodibility factors, numerical model estimates of global mineral dust emissions vary between 1500 and 1800 Tg/yr [16, 43, 44]. Injection heights are usually restricted to the planetary boundary layer (2–4 km) but may reach up to 6 km dependent on meteorological conditions [45]. Although mineral dust is usually considered of natural origin, it is estimated that about 30% of the mineral dust load in the atmosphere could be ascribed to human activities through desertification and land misuse. The Sahara desert is the major source of mineral dust, which subsequently spreads across the Mediterranean and European region, across the Caribbean Sea, and towards Central and North America. Additionally it plays a significant role in the nutrient inflow to the Amazon rainforest [25]. The Gobi Desert is another important source of dust in the atmosphere, which affects eastern Asia and western North America.

Volcanic ash is formed during explosive volcanic eruptions, phreatomagmatic eruptions, or pyroclastic density currents (see Section 2.3.1). On average about 20 volcanoes erupt at any given time worldwide, 50–70 volcanoes erupt throughout a year, and at least one large eruption with a Volcanic Explosivity Index (VEI, relative measure of the explosiveness of volcanic eruption; [46]) greater than 4 occurs annually [47]. The total emissions of volcanic ash into the troposphere by small volcanic eruptions with VEI < 4 (these eruptions make up the majority in number) is estimated to be 20 Tg/yr [48], equivalent to 10 km^3 when assuming a particle density of 2000 kg/m^3. However, these volcanic ash emissions are usually removed from the atmosphere quickly and are therefore only of local interest in the vicinity of the volcanoes up to a distance of about one hundred kilometers. Table 2 summarises ash emissions from major volcanic eruptions, lasting from a few hours to several weeks since 1900 with VEI ≥ 4 with a tephra release ranging from 0.1 to 100 km^3. It is obvious, that stronger volcanic eruptions are generally less frequent [28]. Note that before the satellite era starting about 1980, our knowledge on volcanic eruptions with VEI ≤ 4 is probably not complete due to limited observations of remote volcanic eruptions. Dependent on meteorological conditions and the injection height of the volcanic emissions, volcanic ash from eruptions with VEI ≥ 4 may be transported over thousands of kilometres in the atmosphere [49]. Therefore, volcanic ash, although released sporadically during volcanic eruptions, is an abundant atmospheric species.

Table 2: Tephra mass release in DRE (Dense Rock Equivalent) of well-known volcanic eruptions since 1900 given for VEI values from 4 to 6 in three mass ranges: 0.1–1 km^3, 1–10 km^3, and 10–100 km^3 ([47]

0.1 km^3		1 km^3		10 km^3		100 km^3
VEI = 4		**VEI = 5**		**VEI = 6**		
Nabro 2011 and Puyehue-Cordón Caulle 2011						
Grimsvötn 2011 and Merapi 2010						
Eyjafjallajökull 2010 and Sarychev Peak 2010						
Kasatochi 2008 and Chaiten 2008						

Reventator 2002 and Ulawun 2000					
Lascar 1993 and Mt. Spurr 1992					
Kelud 1990 and Kiluchevkoi 1987		Mt. Hudson 1991		Pinatubo 1991	
Chikurachki 1986 and Mount Augustine 1986					
Colo 1983 and Galunggung 1982		El Chichon 1982			
Pagan 1981 and Alaid 1981					
Mount Augustine 1976 and Tolbachik 1975		Mt. St. Helens 1980			
Volcan de Fuego 1974 and Tiatia 1973					
Fernandina 1968 and Mount Awu 1966					
Kelud 1966 and Taal 1965					
Shiveluch 1964 and Carran-Los Venados 1955		Agung 1963			
Mount Spurr 1953 and Bagana 1952		Bezymianny 1956			
Kelud 1951 and Mount Lamington 1951					
Ambrym 1950 and Hekla 1947					
Sarychev Paek 1946 and Avachinsky 1945					
Paricutin 1943–1952 and Suoh 1933					
Volcan De Fuego 1932 and Mont Aniakchak 1931		Kharimkotan 1933			
Kliucheskoi 1931 and Komagatake 1931		Cerro Azul 1932			
Komagatake 1929 and Avachinsky 1926					
Raikoko 1924 and Manam 1919					
Kelud 1919 and Agrhan 1917		Katla 1918			
Tungurahua 1916 and Sakurajima 1914					
Mount Lolobau 1911 and Grimsvötn 1903		Colima 1913		Katmai/ Novarupta 1912	

Monut Pelee 1902		Ksudach 1907		St. Maria 1902	

It should be considered as well that fresh volcanic ash may be remobilised into the atmosphere from ash deposits on land, similar to what is observed for mineral dust, particularly in arid and semiarid regions. This contradicts with the general assumption that volcanic ash environmental and climate effects are restricted only to the duration of a volcanic eruption with time scales of days to weeks. Reference [9], for example, reports such posteruptive volcanic ash clouds being transported over the Patagonian desert for several months to years after the 1991 eruption of Mt. Hudson in Chile (Figure 2). Following the recent eruption of Eyjafjallajökull in Iceland during April/May 2010, volcanic ash remobilisation created poor air quality and health concerns for the local population for several months [10–12]. References [8, 9] also report volcanic ash from resuspensions events from the Katmai/ Novarupta eruption in 1912, which typically occur in the fall before snowfall. The June 6–8, 1912 Katmai/Novarupta eruption was the largest volcanic eruption in the 20th Century and produced volcanic ash deposits of 1–10 meters around the volcano. A lack of snow combined with strong northerly winds is able to mobilise the hundred years old volcanic ash into the atmosphere even nowadays (Figure 2).

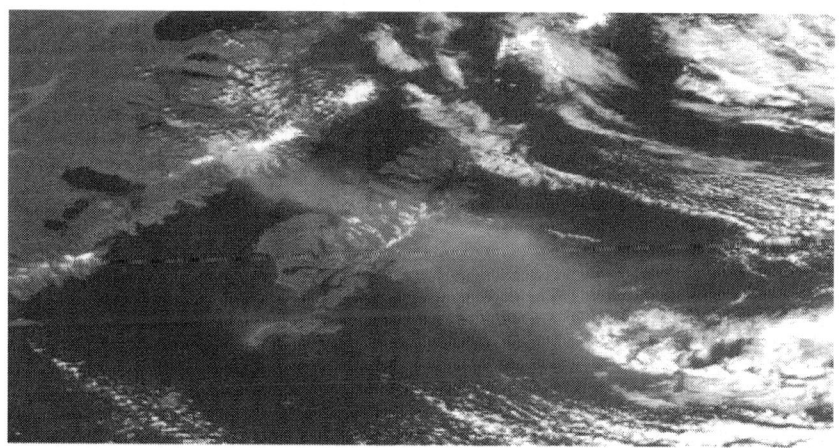

(a)

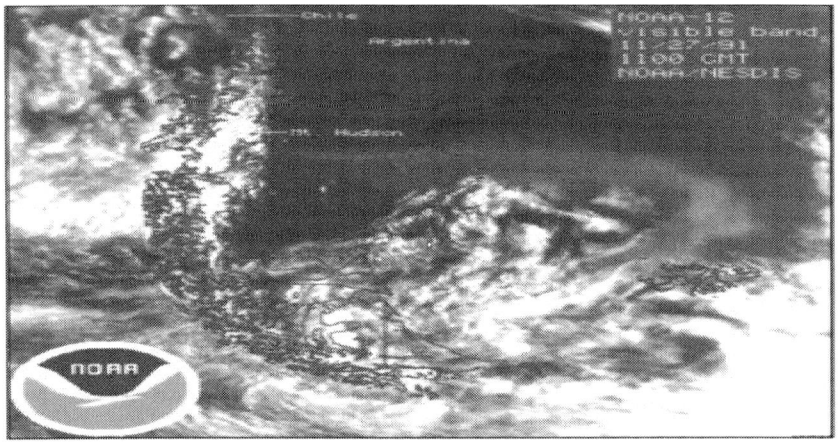

(b)

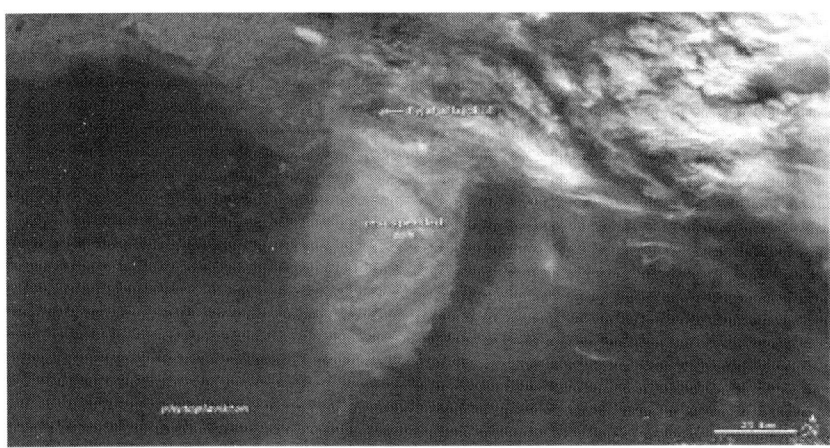

(c)

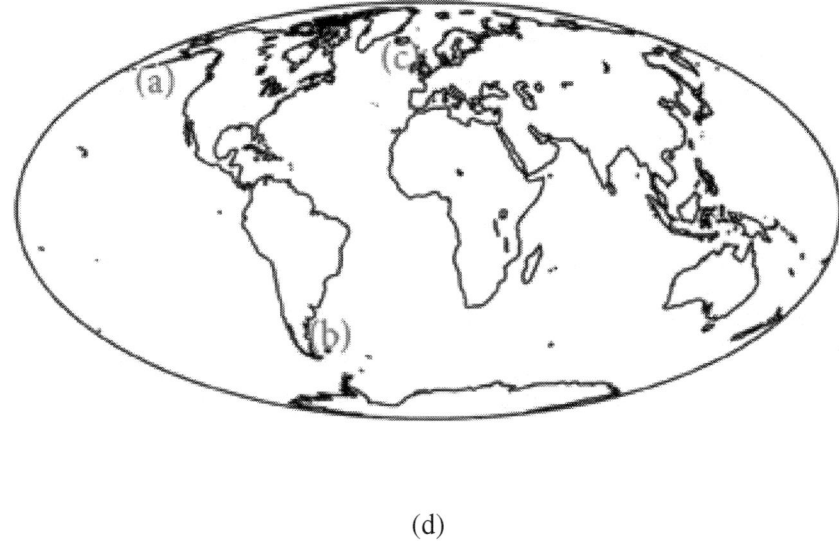

(d)

Figure 2: Volcanic ash resuspension event as seen from satellite: (a) September 21, 2003: Katmai/Novarupta, Alaska; (b) November 27, 1991: Cerro Hudson, Chile; (c) May 27, 2010: Eyafjallajökull, Iceland; (d) approximate location of (a)–(c). Courtesy of NASA.

Atmospheric Load

Atmospheric concentrations of mineral dust and volcanic ash are subject to considerable temporal and spatial variability. Seasonal variability, for example, rainy and dry seasons, determines to a great extent the mineral dust load in the atmosphere, whereas volcanic ash atmospheric load is mainly dependent on the occurrences of sporadic and usually unpredictable volcanic eruptions.

Measurements in the Sahelian belt of West Africa [50] during 2006 to 2008 reveal median daily mineral dust concentrations of around $80\,\mu g/m^3$ with about 40% exceeding $100\,\mu g/m^3$ and less than 3% exceeding $500\,\mu g/m^3$. The maximum measured daily concentrations range is between 2250 and $4020\,\mu g/m^3$ [50]. The European standard for air quality (daily mean PM10 concentration of $50\,\mu g/m^3$ should not be exceeded for more than 35 days per year) is exceeded in this area by mineral dust aerosols about 200 days per year [50]. In the northern part of the Taklamakan desert a higher interannual variability with maximum daily mineral dust concentrations of 645–3800 µg/m^3 was measured between 2001 and 2004 [51]. An even higher variability is reported for the Inner Mongolia region [52] with PM10 concentrations of 190–9625 µg/m^3. Downwind from main mineral dust sources, concentrations are usually smaller, for example, on Cape Verde Islands between 65 and 264 µg/m^3 during mineral dust transport episodes [53].

During the eruption of Eyjafjallajökull on Iceland in 2010, maximum ash concentrations up to 4000 µg/m^3 are reported from measurements of the volcanic ash cloud spreading over Europe [54, 55], exceeding the threshold for safe aviation (2000 µg/m^3). Daily mean near surface concentrations reached up to 400 µg/m^3 in Scandinavia during the eruption [54]. Close to Eyjafjallajökull, the maximum daily average near surface concentration exceeded 1230 µg/m^3 during the ongoing eruption, but also after the eruption stopped, maximum daily average concentration reached more than 1000 µg/m^3 during resuspension events [12]. Volcanic ash concentrations within the volcanic eruption plume are expected to be even higher. However, measurements are difficult to obtain as saturation levels are reached by remote sensing instruments and direct measurements destroy measurement equipment and are too dangerous for humans to approach too close to the eruption. Altogether, the magnitudes of maximum atmospheric concentrations that may be reached during mineral dust storms and volcanic eruptions are relatively similar, although for volcanic eruptions with VEI $\geq$ 4 higher atmospheric ash concentrations are expected.

Deposition

Mineral dust and volcanic ash are removed from the atmosphere by gravitational settling, turbulent dry deposition, and wet scavenging by rain called wet deposition [56]. Dry and wet aggregation of volcanic ash belongs per definition to the removal processes of dry and wet deposition [6]. However, because of their specialty they are discussed separately later. The ratio of dry-to-wet deposition differs considerably in mineral dust model estimates [19], with wet deposition over the ocean ranging from 30% to 95% of the total mineral dust deposition [18]. It should be noted that all uncertainties of the global mineral dust cycle, including emissions (see Section 3.1) and transport in the atmosphere, are summed up in the distribution and amount of deposition fluxes. Mineral dust deposition to ice cores reveals an up to factor 100 difference between cold and warm periods in the geological past [16], for example, much higher deposition during the Last Glacial Maximum, 21,000 years before present, compared to the present-day climate.

Reference [57] estimated the millennial scale flux of volcanic ash into the Pacific Ocean based on the marine sediment core data [58, 59] to be about 128–221 Tg/yr. This represents a conservative estimate, as already one volcanic eruption of VEI = 4 like Kasatochi in 2008 [60] can produce higher deposition fluxes. The estimated millennial volcanic ash flux is comparable to the mineral dust flux into the Pacific Ocean of around 100 Tg/yr [16]. So far, however, the importance of volcanic eruptions on climate, for example, by modifying the biogeochemistry of the surface ocean, has gained limited attention compared to the much better investigated effects of mineral dust. The amount of volcanic ash and bioavailable iron attached to the ash surface deposited into the ocean during episodic large volcanic eruptions may exceed the annual mineral dust flux by far. Reference [61], for example, estimated that iron deposition of the Mt. Hudson's volcanic eruption in Chile during August 12–15, 1991 is equivalent to ~500 years of Patagonian iron dust fallout.

Gravitational settling of volcanic ash has been observed to exceed the terminal settling velocity of single ash particles [6, 62]. This is explained by the formation of aggregates in the volcanic plume, as well as under more diluted conditions in the volcanic cloud [63–65]. Volcanic ash aggregation therefore represents a process that increases sedimentation and reduces atmospheric volcanic ash concentration during long-range transport. However, key aggregate formation processes and basic classifications are topics of ongoing debate. Several important questions remain unanswered up to now. Is aggregation driven primarily by hydrometeor formation? How does aggregation vary in time and space? What is the role of electrostatic charge and "secondary minerals"? How do instabilities (e.g., mammatus) change deposition rates? What proportion of fine-grained ash ends up in aggregates? Where does particle aggregation mainly occur (e.g., vertical plume, horizontal cloud, or during atmospheric sedimentation)? A better understanding of volcanic ash aggregation will be necessary to improve the modelling of volcanic ash dispersion and deposition.

ATMOSPHERIC PROCESSING

Chemical Processing in Volcanic Plumes

Before atmospheric processing occurs at ambient temperatures, volcanic ash undergoes extreme temperature gradients (from about 1000°C to less than 0°C) in extreme short periods of time (few minutes) in the volcanic eruption plume [1, 66, 67], which is expanding vertically into the atmosphere from the vent to the level of neutral buoyancy (Figure 1). Besides fragmentation processes taking place here (see Section 2.3.1), quenching represents an important process in the production of the glass material contained in volcanic ash together with minerals formed by incomplete crystallisation reactions (see Section 2.2). Intensive lightning in the volcanic eruption plume is an often-observed phenomenon [68, 69] due to vertically separated regions of oppositely charged volcanic ash particles. Aggregation processes of volcanic ash (see Section 3.3) may be affected by the charged volcanic ash particles. Until now, the temperature and ionising effects of lighting strokes on volcanic ash chemical composition have not been investigated. A number of potentially important processes for physical-chemical modifications of volcanic ash surfaces, without considering lightning, are discussed in the literature, as summarised below. Volcanic eruption plumes cool significantly during rise from about 1000°C at the vent to ambient temperature leading to various homogeneous and heterogeneous chemical and microphysical modifications on the ash surfaces. During volcanic eruptions, large amounts of volatiles [1, 38,66] are released into the atmosphere along with volcanic ash. Through the interaction between these gases and secondary aerosols produced from these gases with volcanic ash within the eruption plume, it is assumed that soluble compounds are produced on the volcanic ash surfaces [70–72], scavenging up to 30%–40% of the sulfur and 10%–20% of the chlorine released from volcanic eruptions [70, 73]. Reference [71] introduced the idea that scavenging of volatiles by ash within eruption plumes occurs in three temperature-dependent zones: (1) the "salt formation zone" representing the hot core of the eruption plume where sulfate and halide salt aerosols, which were formed at near magmatic temperatures, are adsorbed onto ash particles; (2) in the "surface adsorption zone" halogen gases react directly with the surface of ash during the cooling of the plume until temperatures of about 700°C are reached; (3) the "condensation zone" is characterised by the formation of sulfuric and halogen acids at temperatures below 338°C.

Leaching experiments with pristine volcanic ash in water have been performed for decades [74] revealing the release of various sulfate and halide compounds in addition to biologically relevant elements such as N, P, Si, Fe, Cd, Co, Cu, Mn, Mo, Pb, and Zn [75–77]. Several studies suggest adsorption of volcanic salts on volcanic ash surfaces as the main mechanism for the production of soluble compounds on volcanic ash [78, 79]. Other studies emphasise that condensation of sulfuric acid onto volcanic ash may drive dissolution reactions, thereby providing the source of the soluble cations measured in ash leachates [70]. Different to these assumptions, measurements of [80] point to a rapid acid dissolution of the ash surface material within eruption plumes followed by precipitation of secondary minerals and salts at the ash-liquid interface. Reference [80] assumes adsorption of volcanic salts to represent a process of minor importance. Reference [81] reports about the high-temperature scavenging of volcanic SO_2 by volcanic ash with potential important modifications of the ash surfaces under cooler conditions. Only a few of these aspects have been explored so far by using complex plume models, for example, [82]. Despite the progress made in recent years, the physical-chemical mechanisms, which govern the modification of the surface composition of volcanic ash during its transit through the volcanic eruption plume, where large temperature (from about 1000°C to ambient temperature) and chemical gradients prevail, are poorly understood [83] and require further investigations. A further understanding of these processes under such extreme conditions may also help to increase our knowledge of atmospheric processing of mineral dust, even though mineral dust modifications are restricted to ambient temperatures.

Chemical Processing at Ambient Atmospheric Temperatures

After a volcanic plume reaches neutral buoyancy conditions, the volcanic cloud spreads out more horizontally (Figure 1). In the volcanic cloud further chemical modifications of volcanic ash surfaces at ambient temperatures, for example, during long-range atmospheric transport at high SO_2/sulfate concentrations, may also occur. Although the volcanic emissions are already diluted in the volcanic cloud, the acidic environment in volcanic clouds, which is dependent on the volcanic release rate of SO_2 [84, 85] (Figure 3) and other acidic substances like HCl and HF, may exceed by far that for mineral dust acid processing in anthropogenic polluted air masses [86]. However, volcanic ash, volcanic gases, and their oxidation products (e.g., H_2SO_4) are not necessarily released into similar atmospheric altitudes during a volcanic eruption (e.g., 1991 eruption of Pinatubo; [87]), resulting in different altitudes for the major dispersion pathways of volcanic ash and gases and limited acid processing during atmospheric transport. Furthermore, due to rapid sedimentation, a separation of volcanic ash particles from volcanic gases and secondary aerosols like sulphate occurs within some hours to days (dependent on the volcanic ash size distribution), even when the eruption height for all volcanic emissions is very similar as what was observed during the eruption of Kasatochi in 2008 [49].

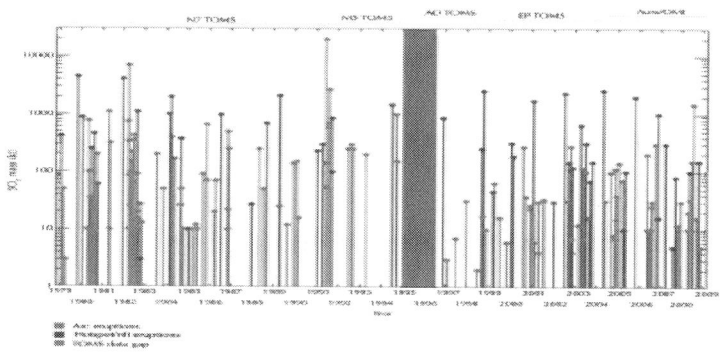

Figure 3: SO_2 emissions from major volcanic eruptions observed from satellite. Arc eruptions with SO_2 emissions exceeding 1000 kt: Mt. St. Helens 1980 (slightly less), Alaid 1981, El Chichon 1982, Mt. Pinatubo 1991, Mt. Hudson 1991, Raboul 1994, and Kasatochi 2008 (Courtesy of NASA).

In contrast, volcanic ash, which is remobilised from ash deposits, can be assumed to undergo very similar atmospheric processing as mineral dust. The importance of photochemistry for mineral dust under atmospheric conditions is highlighted in several studies [22, 23, 88, 89]. Mineral dust particles can act as a sink for SO_2, thus enabling the formation of sulfate on the mineral dust surfaces [90]. When mineral dust concentrations are low, they may trigger the nucleation of new sulfate particles via a series of photochemical reactions involving the mineral dust surface [91]. Metal oxides present in mineral dust act as atmospheric photocatalysts promoting the formation of gaseous OH radicals, which initiate the conversion of SO_2 to H_2SO_4 in the vicinity of dust particles. Comparable results for volcanic ash have also been measured (Dupart, personal information, 2011).

At ambient temperature, when clouds are present in the atmosphere, cloud processing is assumed to provide the main mechanism for the uptake of acid gases in the atmosphere by aerosols in general [56], including mineral dust [92–94] and volcanic ash particles [95]. During atmospheric transport of mineral dust and volcanic ash particles, clouds often evaporate leaving only a thin film of aqueous electrolyte around each particle [56]. This film of aerosol water is very acidic compared to the cloud droplet, reaching pH values of 2 or even lower [86, 96]. As clouds can form and evaporate several times [56], five to ten cycles of pH alternation in the aqueous film around mineral dust and volcanic ash particles can occur before these particles are deposited via wet deposition or sedimentation out of the atmosphere [30]. Atmospheric processing is widely accepted to represent a key process for iron solubility in mineral dust [30, 97, 98]. Limited understanding of these processes hinders the development of accurate biogeochemical models predicting the impact of mineral dust and volcanic ash deposition on the chemistry of trace metals in the surface ocean and ultimately on the global carbon cycle (see Section 5.4.3).

Atmospheric processing of mineral dust and volcanic ash particles at freezing temperatures is a process, which has been rarely studied. However, indications for increased iron solubility are presented [99]. Physical-chemical modifications of volcanic ash surfaces in the volcanic plume and/or volcanic cloud may also affect volcanic ash aggregation (see Section 3.3) [6], which in turn influences particle sedimentation from the plume and during long-range transport, thereby affecting the residence time of volcanic ash in the atmosphere.

Atmospheric and volcanic processing with modifications of the surface chemical composition of mineral dust and volcanic ash particles has implications on their behaviour to act as cloud condensation or ice nuclei (CCN or IN; see Section 5.4.2) and releasing nutrients (in particular iron) on contact with seawater (see Section5.4.3). Observational studies in the field and the laboratory have confirmed that mineral dust and volcanic ash particles can act as CCN or IN, in particular, after being aged in the atmosphere [20, 21, 95, 100–102]. If the surface of volcanic ash has been partially dissolved, water uptake may be favoured [39], thereby explaining the ability of volcanic ash to act as cloud condensation and ice nuclei.

For the sake of completeness, mechanical and biogeochemical weathering that takes place at the Earth's surface under atmospheric conditions is mentioned here as well. These processes are important in the generation of mineral dust and decomposition of volcanic ash on geological time scales [103]. However, they are beyond the scope of this review. The interested reader is referred to the extensive literature, for example, [104], which provides detailed information on weathering processes and important control variables, such as water, oxygen, and acids, as well as lithology, morphology, soil, ecosystems land use, temperature, and runoff.

ENVIRONMENTAL AND CLIMATE IMPACTS

Human Health

Aeolian dust episodes represent a major health concern for humans due to elevated atmospheric concentrations (see Section 3.2). In urban centres in Asia, mineral dust episodes have become a growing concern for the health of populations and ecosystems due to mixing with pollution aerosols [105, 106]. With higher levels of exposure, widespread chronic respiratory and lung diseases, asthma, allergic alveolitis, and eye irritations are well-documented health effects [107]. Human exposure to high volcanic ash concentrations may create similar health effects as those related with mineral dust. However, due to its sharp surface structures, small ash particles have additional mechanical effects; for example, they can abrade the front of the eye under windy conditions [108] (Horwell and Baxter, 2006) and lead to silicosis in the worst case.

Aviation

Strong mineral dust storms mainly affect take-off and landing of aircraft due to poor visibility. However, aircraft can operate in environments with high mineral dust concentration without any engine problems. Mineral dust which typically melts at temperatures of around 1700°C does not melt when it is ingested in jet engines. However, jet engines may have problems with volcanic ash as the melting temperatures are at or below the operating temperatures of high-performance jet engines, which are around 1400°C [109]. The molten material deposited on the cooler parts of the engines can cause flame-outs, which may also result from glass shards when temperature exceeds their glass transition temperature. 129 flights in the last 60 years were affected by volcanic ash [110], for example, during the eruption of Galunggung, Indonesia (1982) or Redoubt, Alsaka (1989/1990). In nine cases, one or more engines temporarily failed due to the melting of ash in the jet engine turbine [110]. To prevent dangerous flights in the presence of volcanic ash, Volcanic Ash Advisory Centres (VAACs) have been installed with the mandate to regularly publish warnings on the location of volcanic ash in the atmosphere. After the eruption of Eyjafjallakökull in Iceland in 2010, a growing demand developed, suggesting that VAACs should not only follow the zero tolerance rule, but also provide absolute volcanic ash concentrations [32, 111], with a preliminary threshold of 2 mg/m^3 for safe aircraft operation.

Soil Fertilisation

Heavy volcanic ash or mineral dust deposition completely buries vegetation and soil. Plant survival is dependent on deposit thickness, chemistry, compaction, rainfall, and duration of burial. Slight deposition of volcanic ash and mineral dust can affect vegetation and soil positively and negatively. Although thin volcanic ash fall inhibits transpiration and photosynthesis and alters growth, buried plants may survive [37]. A positive effect is attributed to mineral dust originating from the Sahara desert, which is regularly transported across the Atlantic Ocean and is argued to represent the main mineral fertiliser of the Amazon region [25]. A positive effect of volcanic ash deposition is an increase in agricultural production due to mulching, for example, following the 1980 Mount St. Helens eruption [112]. Other positive effects are due to the leaching of nutrients released from volcanic ash [28], for example, following the 1995/96 Mt. Ruapehu eruption. In particular, nutrient-poor soils benefit from volcanic ash leaching; however, soluble toxic elements may also be washed-out [37].

Climate

Direct Radiative Effects

Short-term effects during mineral dust storms, volcanic eruptions, and volcanic ash resuspension events considerably reduce visibility and solar irradiation reaching the Earth's surface, whereas long-term effects occur in a more diluted environment. These effects can be measured by an increase in atmospheric optical depth, represented by enhanced absorption and/or scattering of solar and thermal radiation and modifications in surface temperature. During the first two days following the 1980 eruption of Mount St. Helens, [113] reported a decrease in daytime surface temperature of 8°C induced by volcanic ash absorption of solar radiation in the ash affected area downwind of Mt. St. Helens. Here, the colour of volcanic ash pays an important role; low-Si, high-Fe ash tends to develop a dark brown and black colouration, whereas high-Si ash with low Fe ash tends to appear pale and white coloured [37]. The colour of volcanic ash is also an important factor for the modification of the surface albedo for volcanic ash deposits covering the soil. Reference [112] estimated that the light-coloured volcanic ash from Mount St. Helens reflected two to three times more incoming solar radiation than ash-free soils. In contrast, dark-coloured ash will decrease the surface albedo. A modification of the surface albedo will also occur after volcanic ash deposition on snow or ice. In most cases a reduction is observed, although [114] argues that reflectivity measurements of dry volcanic ash can show albedo values as high as snow. Therefore, large areas covered by volcanic ash may have considerable implications for climate [114].

Direct radiative effects of mineral dust have been studied widely [13, 19, 31, 115–117]. Reference [111] compared optical properties of volcanic ash from the Ejyafjallajökull eruption in 2010 with Sahara dust. Although the optical properties of volcanic ash and mineral dust are relatively similar, the imaginary part of the refractive index shows weaker absorption and drops stronger between the blue and red parts of the spectrum for mineral dust when compared with volcanic ash. Another difference in optical properties was observed for the polarisation properties, with slightly stronger forward and lower backward scattering for volcanic ash. These differences allow discrimination between mineral dust and volcanic ash during, for example, lidar measurements [118]. However, atmospheric processing (see Section 4) may modify the radiative properties of both, mineral dust [119] and volcanic ash, although further investigations are needed.

Due to the relatively short residence time of mineral dust and volcanic ash in the atmosphere (in the order of a few days), their direct radiative effects (and indirect radiative effects; see Section 5.4.2) exhibit a high temporal and spatial variability, with major effects around the main source regions and the main transport pathways from a few kilometers to some thousand kilometres. In these regions, however, the direct radiative effects of mineral dust and volcanic ash may be dominant.

Indirect Radiative Effects

With about 60% of the Earth being cloud covered, clouds represent an important factor in regulating the Earth's radiation budget [120]. It has been estimated that a 5% increase of the shortwave cloud forcing could compensate the radiative effect due to increased greenhouse gases between the years 1750–2000 [107]. Cloud formation, lifetime, and radiative properties are affected by aerosols. These so-called indirect aerosol effects are related to changes of the cloud droplet and ice spectrum.

In general, a greater quantity of cloud droplets are formed with typically smaller size if more aerosols are available to act as cloud condensation nuclei (CCN). High cloud droplet number concentrations (CDNC) reduce the diffusional droplet growth. Therefore droplet sizes cannot be reached which are large enough for an efficient growth by droplet collision. Thus, changes in CDNC can influence cloud albedo (first indirect aerosol effect) [121], cloud lifetime, and precipitation formation (second indirect aerosol effect) [122]. In the tropics, the modification of precipitation formation in deep convective clouds is of special importance. The local effect near the aerosol source regions may lead to suppression of precipitation. However, as more liquid water and water vapour stays in the atmosphere, precipitation will be formed elsewhere and the potential of flooding and erosion is increased. Global circulation can be influenced by the release of latent heat due to condensation of water vapour. Therefore, aerosols including mineral dust and volcanic ash, which act as cloud condensation [20, 123] and ice nuclei (see later), have the potential to affect the Earth's radiation budget, the hydrological cycle, and regional or even global circulation [124]. Satellite observations in the South Atlantic and North Pacific [123] show that either natural degassing or weakly explosive volcanoes affect low marine stratocumulus for up to 1300 km downwind by decreasing the effective radius of droplets and increasing visible brightness, which may add cloud cover in otherwise cloudless areas.

Super-cooled clouds are abundant in the atmosphere, which contain metastable water that freezes as soon as suitable ice nuclei are available. In the presence of particulate material, such as mineral dust, volcanic ash, or pollen [95, 100, 125–130], it may undergo heterogeneous nucleation, where freezing may be initiated at significantly lower supersaturations and higher temperatures than in the case of homogeneous freezing [125]. Upon glaciation, the size distribution and the lifetime and radiative forcing of the clouds are modified. A negative correlation between super-cooled clouds (at −20°C) and the occurrence of mineral dust has been found by [131], likely due to glaciation by dust. Reference [100] measured ice formation in the presence of fine volcanic ash particles between about 250 and 260 K. During the Eyjafjallajökull volcanic eruption on Iceland in spring 2010, lidar measurements over Europe clearly observed volcanic ash having an impact on cloud glaciation [130], and in central Germany the highest IN number concentrations within a two year record of daily IN measurements were measured [95]. Also in Israel, about 5000 km away from Eyjafjallajökull, INs were as high in spring 2010 as during desert dust storms. Reference [95] showed that aging increased the ice nucleating ability of the volcanic ash during its transport in the atmosphere (see Section 4.2).

Ocean Fertilisation

In the past 20 years, iron-enrichment experiments ranging from bottle incubations to open-ocean amendment studies in regions of 50–100 km² have demonstrated that iron supply stimulates phytoplankton growth in High-Nutrient-Low-Chlorophyll (HNLC) waters [132]. Thereby, the surface ocean is fertilised with iron, which affects marine primary production (MPP), phytoplankton community structures, and subsequently has an impact on higher trophic levels of the oceanic food-web (zooplankton, fish). Through the conversion of CO_2 to organic carbon and the sinking of parts of this organic matter into the deep ocean, the process referred to as the "biological pump" is activated and atmospheric CO_2 concentrations can be modified. However, it has been difficult to quantify export production via subsurface storage of carbon. The details of the "iron hypothesis" [24] and the possible magnitude of its effect on the global carbon cycle are subject of intense international debate, particularly in connection with climate engineering.

Ocean fertilisation by mineral dust has been studied extensively, as mineral dust has long been assumed to be the main component of atmospheric deposition of minerals into the open ocean [15]. A significant correlation of dust with climate indicators is found in paleorecords such as ice cores [133–135]. In the NE Pacific Ocean, the supply of iron from dust sources occurs episodically, for example, [136, 137], which is dependent on dust storm frequency and atmospheric circulation. For the Southern Ocean, an important multiproxy dataset was recently presented from a marine sediment core in the sub-Antarctic Atlantic [138]. A close correlation was observed between iron input and marine export production, implying that the process of ion fertilisation on marine biota was a recurred process operating in the sub-Antarctic region over the glacial/interglacial cycles of the last 1.1 Ma. A 25%–50% decrease in CO_2 observed during glacial maxima is attributed to mineral dust [132, 139].

Volcanic ash deposition into the ocean represents another external and largely neglected source of iron. However, its significance and impact on climate has long been considered negligible. The major climate forcing effect following volcanic eruptions is widely assumed to occur due to the reduction of solar radiation through volcanic sulfate aerosols [5]. In contrast to volcanic gases and aerosols, volcanic ash is removed from the atmosphere much faster after an eruption. Recent work, however, showed that volcanic ash modifies the biogeochemical processes in the surface ocean [57, 60, 76, 140, 141] thereby directly affecting climate. When airborne volcanic ash is deposited in the surface ocean, it may release trace species upon contact with seawater [75, 142]. Volcanic ash, though released sporadically, can therefore play a similar role as mineral dust. Other trace metals contained in volcanic ash such as zinc or copper may have both, fertilising or toxic effects on phytoplankton [140].

The first direct evidence for iron fertilisation in an HNLC ocean area by volcanic ash emerged after the eruption of the Kasatochi volcano, situated on the Aleutian Islands in August 2008. Atmospheric and oceanic conditions in the NE Pacific were ideal for generating a massive and large-scale phytoplankton bloom, which was observed by satellite instruments [60], confirmed by insitu measurements [143, 144] and ocean biogeochemical modelling [145]. In 2010, it was speculated that the population of sockeye salmon returning to the Fraser River in Canada which was the largest for decades was associated with the fertilisation of the NE Pacific Ocean by Kasatochi ash in 2008 [146]. However, the effect of volcanic ash on salmon populations is discussed controversially; for example, the analysis of [147] rejects the hypothesis of [146].

After the eruption of Kasatochi in 2008 on the Aleutian Islands, atmospheric CO_2 decreased slightly by ~0.01 Pg C as diatoms and mesozooplankton increased export of organic carbon from the surface to the deeper ocean [143]. This carbon sequestration was negligible compared to the rate at which fossil fuel emissions are rising (7–9 Pg C/yr; [107]). While the volcanic ash flux from Kasatochi of 0.2–0.3 km^3 [60] was relatively small, there is abundant evidence for regular volcanic ash emissions into the atmosphere (see Section 3.1, Table 2). Although strong volcanic eruptions with VEI ≥ 5 are rare and not necessarily close to an oceanic HNLC area, they are argued to have affected MPP and atmospheric CO_2 on geological time scales [148–152]. In addition, volcanic ocean fertilisation is not restricted to HNCL areas, as reported by [153] for the Mediterranean Sea or by [141] for the North Atlantic Ocean. However, all these effects are discussed controversially.

Reduced atmospheric CO_2 concentrations were observed in the years following the 1991 Pinatubo eruption [154, 155]. Reference [156] argued that this was the consequence of increased vegetation photosynthesis induced by the presence of a volcanic sulfate aerosol layer in the atmosphere. Notably, [154] Sarmiento (1993) suggested that the atmospheric CO_2 drawdown was the result of ocean fertilisation by Pinatubo ash. While the 1991 Pinatubo eruption released 5-6 km³ of ash (about 30 times the volume of ash emitted by Kasatochi in 2008), a percentage limited amount fell into the iron-limited Southern Ocean. However, the eruption of Mt. Hudson around the same time deposited approximately 1.1 km³ of ash into the iron-limited Atlantic sector of the Southern Ocean [157]. Surprisingly, this ash deposition event has never been evoked to explain the decrease in atmospheric CO_2 concentration. Furthermore, the fertilisation potential of the Mt. Hudson ash deposited in Patagonia (~1.6 km³), which was easily remobilised by the roaring forties during several months after the eruption has never been considered.

Another interesting event is the eruption of Huaynaputina in Peru in 1600, which produced more than 9.6 km³ of volcanic ash [158], which is known to have settled into the tropical Pacific as well as the Southern Ocean, two large HNLC areas. An iron-fertilisation effect could partly explain the 10 ppm decrease in atmospheric CO_2 concentration measured in Antarctic ice cores after 1600 [159].

Ocean iron fertilisation may also affect the climate relevant exchange of trace gases between the ocean and the atmosphere. An increase of the MPP is accompanied by an increased contribution of organic carbon (OC) to submicron marine aerosols [160] and the release of dimethylsulfide (DMS) [161], oxidised to sulfate in the atmosphere. OC and sulfate aerosols can act as efficient cloud condensation nuclei and significantly influence cloud properties via the indirect aerosol effects (see Section 5.4.2; [120]), thereby further cooling the Earth's surface.

FUTURE RESEARCH NEEDS

Although there are substantial differences in the history of mineral dust and volcanic ash particles before they are released into the atmosphere (see Sections 2 and 3), there are on the other hand a number of similarities in atmospheric processing at ambient temperatures (see Section 4.2) and environmental and climate impacts (see Section 5). Therefore, this review tries to trigger a closer cooperation between the research communities studying mineral dust and volcanic ash atmospheric chemical modifications and impacts.

Model parameterisations of volcanic ash remobilisation from its deposits on land build on mineral dust mobilisation schemes [10]. However, as the availability of ash in its deposits is limited, modified approaches will be necessary considering mass conserving parameterisations, where the migration of deposits is also included. Such parameterisations might be of interest for mineral dust researchers as well.

The extreme conditions for multiphase chemistry in volcanic plumes (see Section 4.1) regarding temperature and their associated gradients, acidity, lightning, and particle load represent an obstacle which hindered an overall understanding of the important processes up to now. Despite these difficulties, the multiphase volcanic plume chemistry under extreme conditions, however, offers the possibilities to illuminate processes which might also be important for mineral dust atmospheric chemical processing under less extreme conditions. Here, in particular, the formation of bioavailable iron on mineral dust and volcanic ash surfaces for ocean fertilisation (see Section 5.4.3) is emphasised. Joint experimental and modelling research projects between mineral dust [18] and volcanic ash researchers could substantially increase our incomplete understanding beyond what we know today, particularly from leaching experiments [74]. Although leaching experiments are extremely important for our current knowledge, they could be even more important if standard protocols would be defined and applied to allow a comparison between the experiments conducted at different laboratories [97]. Also particle size distributions and mineralogy for particle diameters substantially smaller than $2\,\mu m$ should be increasingly studied, particularly by volcanological researchers, as these particles are subject to long-range transport.

During a volcanic eruption, ash particles are easily injected into atmospheric regimes where freezing temperatures prevail, and therefore a better understanding of the processes affected by freezing temperatures, like IN formations or Fe mobilisation and their climate impacts [99], should be studied more systematically. CCN and IN formation is linked with wet deposition processes of mineral dust and volcanic ash out of the atmosphere. Regarding volcanic ash, an improved knowledge of aggregation (see Section 3.3), a deposition process reducing the amount of volcanic ash for long-range transport, is urgently needed. This process is insufficiently handled in all ash dispersion models [162]. Reference [62] requests continued interaction between the meteorological and volcanological communities to achieve advances in understanding the fundamentals of ash aggregation. Besides ash-ice aggregation processes, which considerably increase the terminal settling velocity of the aggregates in comparison to the single fine ash particle and thereby increase the removal rate of volcanic ash, the process of wet deposition of volcanic ash must also be considered as a nonlinear interaction process between volcanic ash and meteorological clouds. Even without the effects of volcanic ash, understanding of the fundamentals of cloud formation is challenging for atmospheric scientists. Interactions of aerosols, including mineral dust, with water and ice in atmospheric clouds and their influence on cloud formation, lifetime, and precipitation formation is one of the hot topics in climate research [120].

For paleoclimate research, the results from terrestrial and marine environmental archives, namely, ice, peat, sea, and ocean sediment cores for mineral dust [139] and volcanic ash deposition [58, 59], need to be assembled to better assess the climate impacts of volcanic ash versus mineral dust during the geological past. However, until we have a good understanding of present day processes, we will not be able to adequately address these processes either in the palaeo-records or with regard to the future impacts of mineral dust in contrast to volcanic ash on climate.

ACKNOWLEDGMENTS

The financial support through the Cluster of Excellence "CliSAP" (EXC177), University of Hamburg, funded through the German Science Foundation (DFG) is gratefully acknowledged. The author thanks Michael Hemming, Gholamali Hoshyaripour, and Matthias Hort for their comments on the paper.

REFERENCES

1. R. Sparks, M. Bursik, J. Gilbert, L. Glaze, H. Sigurdsson, and A. Woods, Volcanic Plumes, John Wiley, Chichester, UK, 1997.

2. W. I. Rose and A. J. Durant, "Fine ash content of explosive eruptions," Journal of Volcanology and Geothermal Research, vol. 186, no. 1-2, pp. 32–39, 2009. · ·

3. D. B. Dingwell, Y. Lavallée, and U. Kueppers, "Volcanic ash: a primary agent in the Earth system,"Physics and Chemistry of the Earth, vol. 45-46, pp. 2–4, 2012. · ·

4. B. Zimanowski, K. Wohletz, P. Dellino, and R. Büttner, "The volcanic ash problem," Journal of Volcanology and Geothermal Research, vol. 122, no. 1-2, pp. 1–5, 2003.

5. A. Robock, "Volcanic eruptions and climate," Reviews of Geophysics, vol. 38, no. 2, pp. 191–219, 2000. · ·

6. R. J. Brown, C. Bonadonna, and A. J. Durant, "A review of volcanic ash aggregation," Physics and Chemistry of the Earth, vol. 45-46, pp. 65–78, 2012. · ·

7. P. V. Hobbs, D. A. Hegg, and L. F. Radke, "Resuspension of volcanic ash from Mount St. Helens,"Journal of Geophysical Research, vol. 88, no. 6, pp. 3919–3921, 1983.

8. D. Hadley, G. L. Hufford, and J. J. Simpson, "Resuspension of relic volcanic ash and dust from katmai: still an aviation hazard," Weather and Forecasting, vol. 19, pp. 829–840, 2004.

9. T. M. Wilson, J. W. Cole, C. Stewart, S. J. Cronin, and D. M. Johnston, "Ash storms: impacts of wind-remobilised volcanic ash on rural communities and agriculture following the 1991 Hudson eruption, southern Patagonia, Chile," Bulletin of Volcanology, vol. 73, no. 3, pp. 223–239, 2011. · ·

10. S. J. Leadbetter, M. C. Hort, S. Von Lwis, K. Weber, and C. S. Witham, "Modeling the resuspension of ash deposited during the eruption of Eyjafjallajökull in spring 2010," Journal of Geophysical Research, vol. 117, no. D20, 2012. · ·

11. T. Thorsteinsson, G. Gísladóttir, J. Bullard, and G. McTainsh, "Dust storm contributions to airborne particulate matter in Reykjavík, Iceland," Atmospheric Environment, vol. 45, no. 32, pp. 5924–5933, 2011. · ·

12. T. Thorsteinsson, T. Jóhannsson, A. Stohl, and N. I. Kristiansen, "High levels of particulate matter in Iceland due to direct ash emissions by the Eyjafjallajökull eruption and resuspension of deposited ash,"Journal of Geophysical Research B, vol. 117, no. B9, 2012. · ·

13. I. Tegen and I. Fung, "Modeling of mineral dust in the atmosphere: sources, transport, and optical thickness," Journal of Geophysical Research, vol. 99, no. D11, pp. 22897–22914, 1994.

14. J. M. Prospero, P. Ginoux, O. Torres, S. E. Nicholson, and T. E. Gill, "Environmental characterization of global sources of atmospheric soil dust identified with the Nimbus 7 Total Ozone Mapping Spectrometer (TOMS) absorbing aerosol product," Reviews of Geophysics, vol. 40, no. 1, pp. 2-1–2-31, 2002.

15. T. D. Jickells, Z. S. An, K. K. Andersen et al., "Global iron connections between desert dust, ocean biogeochemistry, and climate," Science, vol. 308, no. 5718, pp. 67–71, 2005. · ·

16. N. M. Mahowald, A. R. Baker, G. Bergametti et al., "Atmospheric global dust cycle and iron inputs to the ocean," Global Biogeochemical Cycles, vol. 19, no. 4, 2005.

17. Y. Shao, K.-H. Wyrwoll, A. Chappell et al., "Dust cycle: an emerging core theme in Earth system science," Aeolian Research, vol. 2, no. 4, pp. 181–204, 2011. · ·

18. M. Schulz, J. M. Prospero, A. R. Baker et al., "Atmospheric transport and deposition of mineral dust to the ocean: implications for research need," Environmental Science and Technology, vol. 46, pp. 10390–10404, 2012.

19. N. Huneeus, M. Schulz, Y. Balkanski et al., "Global dust model intercomparison in AeroCom phase I,"Atmospheric Chemistry and Physics, vol. 11, no. 15, pp. 7781–7816, 2011. · ·

20. D. Rosenfeld, Y. Rudich, and R. Lahav, "Desert dust suppressing precipitation: a possible desertification feedback loop," Proceedings of the National Academy of Sciences of the United States of America, vol. 98, no. 11, pp. 5975–5980, 2001. · ·

21. P. Kumar, I. N. Sokolik, and A. Nenes, "Measurements of cloud condensation nuclei activity and droplet activation kinetics of fresh unprocessed regional dust samples and minerals,"

Atmospheric Chemistry and Physics, vol. 11, no. 7, pp. 3527–3541, 2011. · ·

22. F. J. Dentener, G. R. Carmichael, Y. Zhang, J. Lelieveld, and P. J. Crutzen, "Role of mineral aerosol as a reactive surface in the global troposphere," Journal of Geophysical Research, vol. 101, no. D17, pp. 22869–22889, 1996. · ·

23. S. E. Bauer, Y. Balkanski, M. Schulz, D. A. Hauglustine, and F. Dentener, "Global modeling of heterogenous chemistry on mineral aerosol surfaces: influence on tropospheric ozone chemistry and comparison to observations," Journal of Geophysical Research, vol. 109, no. 2, p. D2, 2004. · ·

24. J. H. Martin, "Glacial-interglacial CO_2 change: the iron hypothesis," Paleoceanography, vol. 5, no. 1, pp. 1–13, 1990.

25. I. Koren, Y. J. Kaufman, R. Washington et al., "The Bodélé depression: a single spot in the Sahara that provides most of the mineral dust to the Amazon forest," Environmental Research Letters, vol. 1, no. 1, Article ID 014005, 2006. · ·

26. J. G. Calvert, Glossary of Atmospheric Chemistry Terms, IUPAC, 1990.

27. G. Heiken, "Morphology and Petrography of volcanic ashes," Geological Society of America Bulletin, vol. 83, pp. 1961–1988, 1972.

28. H. U. Schmincke, Volcanism, Springer, Berlin, Germany, 2004.

29. M. Nakagawa and T. Ohba, "Minerals in volcanic ash 1: primary minerals and volcanic glass," Global Environmental Research, vol. 6, pp. 41–51, 2003.

30. Z. Shi, M. D. Krom, T. D. Jickels et al., "Impacts on iron solubility in the mineral dust by processes in the source region and the atmosphere: a review," Aeolian Research, vol. 5, pp. 21–42, 2012.

31. P. Formenti, L. Schuetz, Y. Balkanski et al., "Recent progress in understanding physical and chemical properties of African and Asian mineral dust," Atmospheric Chemistry and Physics, vol. 11, pp. 8231–8256, 2011.

32. B. Langmann, A. Folch, M. Hensch, and V. Matthias, "Volcanic ash over Europe during the eruption of Eyjafjallajökull on Iceland,

April–May 2010," Atmospheric Environment, vol. 48, pp. 1–8, 2012. · ·

33.   S. A. Colgate and T. Sigurgeirsson, "Dynamic mixing of water and lava," Nature, vol. 244, no. 5418, pp. 552–555, 1973. · ·

34.   S. Dartevelle, G. G. J. Ernst, and A. Bernard, "Origin of the Mount Pinatubo climactic eruption cloud: implications for volcanic hazards and atmospheric impacts," Geology, vol. 30, no. 7, pp. 663–666, 2002.

35.   Y. H. Lee, K. Chen, and P. J. Adams, "Development of a global model of mineral dust aerosol microphysics," Atmospheric Chemistry and Physics, vol. 9, no. 7, pp. 2441–2458, 2009.

36.   T. M. Wilson, C. Stewart, V. Sword-Daniels et al., "Volcanic ash impacts on critical infrastructure,"Physics and Chemistry of the Earth, vol. 45-46, no. 5, 23 pages, 2012.

37.   P. M. Ayris and P. Delmelle, "The immediate environmental effects of tephra emission," Bulletin of Volcanology, vol. 74, pp. 1905–1936, 2012.

38.   R. B. Symonds, W. I. Rose, G. J. S. Bluth, and T. M. Gerlach, "Volcanic gas studies: methods, results and applications," in Volatiles in Magma, M. R. Caroll and J. R. Holloway, Eds., vol. 30, pp. 1–66, Reviews in Mineralogy and Geochemistry, 1994.

39.   P. Delmelle, F. Villiéras, and M. Pelletier, "Surface area, porosity and water adsorption properties of fine volcanic ash particles," Bulletin of Volcanology, vol. 67, no. 2, pp. 160–169, 2005. · ·

40.   S. Joussaume, "3-dimensional simulations of the atmospheric cycle of desert dust particles using a general-circulation model," Journal of Geophysical Research, vol. 95, pp. 1909–1941, 1990. ·

41.   N. Mahowald, K. Kohfeld, M. Hansson et al., "Dust sources and deposition during the last glacial maximum and current climate: a comparison of model results with paleodata from ice cores and marine sediments," Journal of Geophysical Research, vol. 104, no. D13, pp. 15895–15916, 1999.

42.   B. Marticorena and G. Bergametti, "Modeling the atmospheric dust cycle: 1. Design of a soil-derived dust emission scheme," Journal of Geophysical Research, vol. 100, no. 8, pp. 16415–16430, 1995.

43. I. Tegen, M. Werner, S. P. Harrison, and K. E. Kohfeld, "Relative importance of climate and land use in determining present and future global soil dust emission," Geophysical Research Letters, vol. 31, no. 5, 2004. ··

44. C. S. Zender, H. Bian, and D. Newman, "Mineral Dust Entrainment and Deposition (DEAD) model: description and 1990s dust climatology," Journal of Geophysical Research, vol. 108, no. D14, 2003.

45. P. Formenti, J. L. Rajot, K. Desboeufs et al., "Airborne observations of mineral dust over western Africa in the summer Monsoon season: spatial and vertical variability of physico-chemical and optical properties," Atmospheric Chemistry and Physics, vol. 11, no. 13, pp. 6387–6410, 2011. ··

46. C. G. Newhall and S. Self, "The volcanic explosivity index (VEI): an estimate of explosive magnitude for historical volcanism," Journal of Geophysical Research, vol. 87, no. C2, pp. 1231–1238, 1982.

47. T. Simkin and I. Siebert, Volcanoes of the World, Smithonian Institution; Geoscience Press, Misoula, Mont, USA, 1994.

48. T. A. Mather, D. M. Pyle, and C. Oppenheimer, "Tropospheric volcanic aerosol," in Volcanism and the Earth's Atmosphere, Geophysical Monograph, vol. 139, pp. 89–211, 2003.

49. B. Langmann, K. Zaksek, and M. Hort, "Atmospheric distribution and removal of volcanic ash after the eruption of Kasatochi volcano: a regional model study," Journal of Geophysical Research, vol. 115, no. D2, 2010. ·

50. B. Marticorena, B. Chatenet, J. L. Rajot et al., "Temporal variability of mineral dust concentrations over West Africa: analyses of a pluriannual monitoring from the AMMA Sahelian Dust Transect,"Atmospheric Chemistry and Physics, vol. 10, no. 18, pp. 8899–8915, 2010. ··

51. M. Mikami, O. Abe, M. Du et al., "The impact of Aeolian dust on climate: sino-Japanese cooperative project ADEC," Journal of Arid Land Studies, vol. 11, pp. 211–222, 2002.

52. C. Hoffmann, R. Funk, M. Sommer, and Y. Li, "Temporal variations in PM10 and particle size distribution during Asian dust storms in Inner Mongolia," Atmospheric Environment, vol. 42, no. 36, pp. 8422–8431, 2008. ··

53. I. Chiapello, "Origins of African dust transported over the northeastern tropical Atlantic," Journal of Geophysical Research, vol. 102, no. D12, pp. 13701–13709, 1997.

54. A. J. Prata and A. T. Prata, "Eyjafjallajökull volcanic ash concentrations determined using Spin enhanced visible and infrared imager measurements," Journal of Geophysical Research, vol. 117, no. D20, 2012. ·

55. H. N. Webster, D. J. Thomson, B. T. Johnson et al., "Operational prediction of ash concentrations in the distal volcanic cloud from the 2010 Eyjafjallajökull eruption," Journal of Geophysical Research D, vol. 117, no. D20, 2012. · ·

56. J. H. Seinfeld and S. N. Pandis, Atmospheric Chemistry and Physics: From Air Pollution to Climate Change, John Wiley, New York, NY, USA, 2006.

57. N. Olgun, S. Duggen, P. L. Croot et al., "Surface ocean iron fertilization: the role of airborne volcanic ash from subduction zone and hot spot volcanoes and related iron fluxes into the Pacific Ocean," Global Biogeochemical Cycles, vol. 25, no. 4, 2011.

58. S. M. Straub and H. U. Schmincke, "Evaluating the tephra input into Pacific Ocean sediments: distribution in space and time," Geologische Rundschau, vol. 87, no. 3, pp. 461–476, 1998. · ·

59. S. Kutterolf, A. Freundt, U. Schacht et al., "Pacific offshore record of plinian arc volcanism in Central America: 3. Application to forearc geology," Geochemistry, Geophysics, Geosystems, vol. 9, no. 2, 2008. · ·

60. B. Langmann, K. Zakšek, M. Hort, and S. Duggen, "Volcanic ash as fertiliser for the surface ocean," Atmospheric Chemistry and Physics, vol. 10, no. 8, pp. 3891–3899, 2010.

61. D. M. Gaiero, J.-L. Probst, P. J. Depetris, S. M. Bidart, and L. Leleyter, "Iron and other transition metals in Patagonian riverborne and windborne materials: geochemical control and transport to the southern South Atlantic Ocean," Geochimica et Cosmochimica Acta, vol. 67, no. 19, pp. 3603–3623, 2003. · ·

62. W. I. Rose and A. J. Durant, "Fate of volcanic ash: aggregation and fallout," Geology, vol. 39, no. 9, pp. 895–896, 2011. · ·

63. A. J. Durant, W. I. Rose, A. M. Sarna-Wojcicki, S. Carey, and A. C. M. Volentik, "Hydrometeor-enhanced tephra sedimentation: constraints from the 18 May 1980 eruption of Mount St. Helens,"Journal of Geophysical Research, vol. 114, no. B3, 2009.

64. C. Bonadonna, R. Genco, M. Gouhier et al., "Tephra sedimentation during the 2010 Eyjafjallajökull eruption (Iceland) from deposit, radar, and satellite observations," Journal of Geophysical Research, vol. 116, no. B12, 2011. · ·

65. A. Folch, "A review of tephra transport and dispersal models: evolution, current status and future perspectives," Journal of Volcanology and Geothermal Research, vol. 235, pp. 96–115, 2012.

66. C. Textor, H. F. Graf, C. Timmreck, and A. Robock, "Emissions from volcanoes," in Emissions of Chemical Compounds and Aerosols in the Atmosphere, C. Granier, C. Reeves, and P. Artaxo, Eds., vol. 18 of Advances in Global Change Research, pp. 269–303, Kluwer, Dordrecht, The Netherlands, 2004.

67. L. G. Mastin, "A user-friendly one-dimensional model for wet volcanic plumes," Geochemistry, Geophysics, Geosystems, vol. 8, no. 3, 2007. · ·

68. M. R. James, L. Wilson, S. J. Lane et al., "Electrical charging of volcanic plumes," Space Science Reviews, vol. 137, no. 1–4, pp. 399–418, 2008. · ·

69. S. R. McNutt and E. R. Williams, "Volcanic lightning: global observations and constraints on source mechanisms," Bulletin of Volcanology, vol. 72, no. 10, pp. 1153–1167, 2010. · ·

70. W. I. Rose, "Scavenging of volcanic aerosol by ash: atmospheric and volcanologic implications," Geology, vol. 5, pp. 621–624, 1977.

71. N. Oskarsson, "The interaction between volcanic gases and tephra: fluorine adhering to tephra of the 1970 Hekla eruption," Journal of Volcanology & Geothermal Research, vol. 8, no. 2–4, pp. 251–266, 1980.

72. E. Bagnato, A. Aiuppa, A. Bertagnini et al., "Scavenging of sulphur, halogens and trace metals by volcanic ash: the 2010 Eyjafjallajökull eruption," Geochimica Et Cosmochimica Acta, vol. 103, pp. 138–160, 2013.

73. J. M. de Moor, T. P. Fischer, D. R. Hilton, E. Hauri, L. A. Jaffe, and J. T. Camacho, "Degassing at Anatahan volcano during the May 2003 eruption: implications from petrology, ash leachates, and SO_2 emissions," Journal of Volcanology and Geothermal Research, vol. 146, no. 1–3, pp. 117–138, 2005. · ·

74. C. S. Witham, C. Oppenheimer, and C. J. Horwell, "Volcanic ash-leachates: a review and recommendations for sampling methods," Journal of Volcanology and Geothermal Research, vol. 141, no. 3-4, pp. 299–326, 2005. · ·

75. P. Frogner, S. R. Gíslason, and N. Óskarsson, "Fertilizing potential of volcanic ash in ocean surface water," Geology, vol. 29, no. 6, pp. 487–490, 2001.

76. S. Duggen, P. Croot, U. Schacht, and L. Hoffmann, "Subduction zone volcanic ash can fertilize the surface ocean and stimulate phytoplankton growth: evidence from biogeochemical experiments and satellite data," Geophysical Research Letters, vol. 34, no. 1, Article ID L01612, 2007. · ·

77. S. Duggen, N. Olgun, P. Croot et al., "The role of airborne volcanic ash for the surface ocean biogeochemical iron-cycle: a review," Biogeosciences, vol. 7, no. 3, pp. 827–844, 2010.

78. P. S. Taylor and R. E. Stoiber, "Soluble material on ash from active Central American volcanoes," Geological Society of America Bulletin, vol. 84, pp. 1031–1042, 1973.

79. D. B. Smith, R. A. Zielinski, W. I. Rose Jr., and B. J. Huebert, "Water-soluble material on aerosols collected within volcanic eruption clouds (Fuego, Pacaya, Santiaguito, Guatamala)," Journal of Geophysical Research, vol. 87, no. 7, pp. 4963–4972, 1982.

80. P. Delmelle, M. Lambert, Y. Dufrêne, P. Gerin, and N. Óskarsson, "Gas/aerosol-ash interaction in volcanic plumes: new insights from surface analyses of fine ash particles," Earth and Planetary Science Letters, vol. 259, no. 1-2, pp. 159–170, 2007.

81. P. M. Ayris, A. F. Lee, K. Wilson, U. Kueppers, D. B. Dingwell, and P. Delmelle, "SO_2 sequestration in large volcanic eruptions: high-temperature scavenging by tephra," Geochimica Et Cosmochimica Acta, vol. 110, pp. 58–69, 2013.

82. C. Textor, H. F. Graf, M. Herzog, J. M. Oberhuber, W. I. Rose, and G. G. J. Ernst, "Volcanic particle aggregation in explosive eruption columns. Part I: parameterization of the microphysics of hydrometeors and ash," Journal of Volcanology and Geothermal Research, vol. 150, no. 4, pp. 359–377, 2006. · ·

83. P. Ayris and P. Delmelle, "Volcanic and atmospheric controls on ash iron solubility: a review," Physics and Chemistry of the Earth, vol. 45-46, pp. 103–112, 2012. · ·

84. M. M. Halmer, H.-U. Schmincke, and H.-F. Graf, "The annual volcanic gas input into the atmosphere, in particular into the stratosphere: a global data set for the past 100 years," Journal of Volcanology and Geothermal Research, vol. 115, no. 3-4, pp. 511–528, 2002. · ·

85. C. Gao, A. Robock, and C. Ammann, "Volcanic forcing of climate over the past 1500 years: an improved ice core-based index for climate models," Journal of Geophysical Research, vol. 113, no. D23, 2008. · ·

86. N. Meskhidze, W. L. Chameides, A. Nenes, and G. Chen, "Iron mobilization in mineral dust: can anthropogenic SO_2 emissions affect ocean productivity?" Geophysical Research Letters, vol. 30, no. 21, 2003. · ·

87. J. Fero, S. N. Carey, and J. T. Merrill, "Simulating the dispersal of tephra from the 1991 Pinatubo eruption: implications for the formation of widespread ash layers," Journal of Volcanology and Geothermal Research, vol. 186, no. 1-2, pp. 120–131, 2009. · ·

88. M. Ullerstam, R. Vogt, S. Langer, and E. Ljungström, "The kinetics and mechanism of SO_2 oxidation by O_3 on mineral dust," Physical Chemistry Chemical Physics, vol. 4, no. 19, pp. 4694–4699, 2002. · ·

89. M. Ndour, P. Conchon, B. D'Anna, O. Ka, and C. George, "Photochemistry of mineral dust surface as a potential atmospheric renoxification process," Geophysical Research Letters, vol. 36, no. 5, 2009. · ·

90. R. C. Sullivan, S. A. Guazzotti, D. A. Sodeman, and K. A. Prather, "Direct observations of the atmospheric processing of Asian mineral dust," Atmospheric Chemistry and Physics, vol. 7, no. 5, pp. 1213–1236, 2007.

91. Y. Dupart, S. M. King, B. Nekat et al., "Mineral dust photochemistry induces nucleation events in the presence of SO_2," Proceedings of the National Academy of Science of the United States of America, vol. 109, pp. 20842–20847, 2012.

92. S. Wurzler, T. G. Reisin, and Z. Levin, "Modification of mineral dust particles by cloud processing and subsequent effects on drop size distributions," Journal of Geophysical Research D, vol. 105, no. D4, pp. 4501–4512, 2000.

93. D. S. Mackie, P. W. Boyd, K. A. Hunter, and G. H. McTainsh, "Simulating the cloud processing of iron in Australian dust: pH and dust concentration," Geophysical Research Letters, vol. 32, no. 6, 2005. · ·

94. A. Matsuki, A. Schwarzenboeck, H. Venzac, P. Laj, S. Crumeyrolle, and L. Gomes, "Cloud processing of mineral dust: direct comparison of cloud residual and clear sky particles during AMMA aircraft campaign in summer 2006," Atmospheric Chemistry and Physics, vol. 10, no. 3, pp. 1057–1069, 2010.

95. H. Bingemer, H. Klein, M. Ebert et al., "Atmospheric ice nuclei in the Eyjafjallajökull volcanic ash plume," Atmospheric Chemistry and Physics, vol. 12, pp. 857–867, 2012. · ·

96. F. Solmon, P. Y. Chuang, N. Meskhidze, and Y. Chen, "Acidic processing of mineral dust iron by anthropogenic compounds over the north Pacific Ocean," Journal of Geophysical Research, vol. 114, no. D2, 2009. · ·

97. A. R. Baker and P. L. Croot, "Atmospheric and marine controls on aerosol iron solubility in seawater,"Marine Chemistry, vol. 120, no. 1–4, pp. 4–13, 2010. · ·

98. N. Meskhidze, W. L. Chameides, and A. Nenes, "Dust and pollution: a recipe for enhanced ocean fertilization?" Journal of Geophysical Research D, vol. 110, no. D3, 2005. · ·

99. D. Jeong, K. Kim, and W. Choi, "Accelerated dissolution of iron oxides in ice," Atmospheric Chemistry and Physics, vol. 12, pp. 11125–11133, 2012.

100. A. J. Durant, R. A. Shaw, W. I. Rose, Y. Mi, and G. G. J. Ernst, "Ice nucleation and overseeding of ice in volcanic clouds," Journal of Geophysical Research, vol. 113, no. D9, 2008. · ·

101. Z. Shi, D. Zhang, M. Hayashi, H. Ogata, H. Ji, and W. Fujiie, "Influences of sulfate and nitrate on the hygroscopic behaviour of coarse dust particles," Atmospheric Environment, vol. 42, no. 4, pp. 822–827, 2008. · ·

102. T. L. Lathem, P. Kumar, A. Nenes et al., "Hygroscopic properties of volcanic ash," Geophysical Research Letters, vol. 38, no. 11, Article ID L11802, 2011. · ·

103. R. A. Dahlgren, F. C. Uoolim, and W. H. Casey, "Field weathering rates of Mt. St. Helens tephra,"Geochimica et Cosmochimica Acta, vol. 63, no. 5, pp. 587–598, 1999. · ·

104. A. F. White and S. L. Brantley, "Chemical weathering rates of silicate minerals: an overview," Reviews in Mineralogy and Geochemistry, vol. 31, pp. 1–22, 1995.

105. J.-I. Jeong and S.-U. Park, "Interaction of gaseous pollutants with aerosols in Asia during March 2002,"Science of the Total Environment, vol. 392, no. 2-3, pp. 262–276, 2008. · ·

106. H. Yuan, G. Zhuang, J. Li, Z. Wang, and J. Li, "Mixing of mineral with pollution aerosols in dust season in Beijing: revealed by source apportionment study," Atmospheric Environment, vol. 42, no. 9, pp. 2141–2157, 2008. · ·

107. IPCC (International Panel of Climate Change), Fourth Assessment Report: Climate Change, 2007.

108. C. J. Horwell and P. J. Baxter, "The respiratory health hazards of volcanic ash: a review for volcanic risk mitigation," Bulletin of Volcanology, vol. 69, no. 1, pp. 1–24, 2006. · ·

109. M. G. Dunn, A. J. Baran, and J. Miatech, "Operation of gas turbine engines in volcanic ash clouds,"Journal of Engineering for Gas Turbines and Power, vol. 118, no. 4, pp. 724–731, 1996.

110. M. Guffanti, T. J. Casadevall, and K. Budding, "1953–2009: Encounters of Aircraft with Volcanic Ash Clouds, A Compilation of Known Incidents," U.S. Geological Survey Data Series 545, 12 p., Plus 4 Appendixes Including the Compilation Database, 2010, http://pubs.usgs.gov/ds/545/.

111. B. Weinzierl, D. Sauer, A. Minikin et al., "On the visibility of airborne volcanic ash and mineral dust from the pilot's perspective in flight," Physics and Chemistry of the Earth, vol. 45-46, pp. 87–102, 2012.

112. R. J. Cook, J. C. Barron, R. I. Papendick, and G. J. Williams III, "Impact on agriculture of the Mount St. Helens eruptions," Science, vol. 211, no. 4477, pp. 16–22, 1981.

113. C. Mass and A. Robock, "The short-term influence of the Mount St. Helens volcanic eruption on surface temperature in the northwest United States (Idaho, Montana)," Monthly Weather Review, vol. 110, no. 6, pp. 614–622, 1982.

114. M. T. Jones, R. S. J. Sparks, and P. J. Valdes, "The climatic impact of supervolcanic ash blankets,"Climate Dynamics, vol. 29, no. 6, pp. 553–564, 2007. · ·

115. Y. J. Kaufman, D. Tanré, O. Dubovik, A. Karnieli, and L. A. Remer, "Absorption of sunlight by dust as inferred from satellite and ground-based remote sensing," Geophysical Research Letters, vol. 28, no. 8, pp. 1479–1482, 2001. · ·

116. H. E. Redmond, K. D. Dial, and J. E. Thompson, "Light scattering and absorption by wind blown dust: theory, measurement, and recent data," Aeolian Research, vol. 2, no. 1, pp. 5–26, 2010. · ·

117. J. S. Reid, H. H. Jonsson, H. B. Maring et al., "Comparison of size and morphological measurements of coarse mode dust particles from Africa," Journal of Geophysical Research, vol. 108, no. D19, 2003. · ·

118. S. Groß, V. Freudenthaler, M. Wiegner, J. Gasteiger, A. Geiß, and F. Schnell, "Dual-wavelength linear depolarization ratio of volcanic aerosols: lidar measurements of the Eyjafjallajökull plume over Maisach, Germany," Atmospheric Environment, vol. 48, pp. 85–96, 2012. · ·

119. S. E. Bauer, M. I. Mishchenko, A. A. Lacis, S. Zhang, J. Perlwitz, and S. M. Metzger, "Do sulfate and nitrate coatings on mineral dust have important effects on radiative properties and climate modeling?"Journal of Geophysical Research, vol. 112, no. D6, 2007. · ·

120. U. Lohmann and J. Feichter, "Global indirect aerosol effects: a review," Atmospheric Chemistry and Physics, vol. 5, no. 3, pp. 715–737, 2005.

121. S. Twomey, "The influence of pollution on the shortwave albedo of clouds," Journal of Atmospheric Sciences, vol. 34, pp. 1149–1152, 1977.

122. B. A. Albrecht, "Aerosols, cloud microphysics, and fractional cloudiness," Science, vol. 245, no. 4923, pp. 1227–1230, 1989.

123. S. Gasso, "Satellite observations of the impact of weak volcanic activity on marine clouds," Journal of Geophysical Research, vol. 113, no. D14, 2008. ·

124. F. J. Nober, H.-F. Graf, and D. Rosenfeld, "Sensitivity of the global circulation to the suppression of precipitation by anthropogenic aerosols," Global and Planetary Change, vol. 37, no. 1-2, pp. 57–80, 2003. · ·

125. H. R. Pruppacher and J. D. Klett, Microphysics of Clouds and Precipitation, Kluwer, Dordrecht, The Netherlands, 2nd edition, 1997.

126. W. Szyrmer and I. Zawadzki, "Biogenic and anthropogenic sources of ice-forming nuclei: a review,"Bulletin of the American Meteorological Society, vol. 78, no. 2, pp. 209–228, 1997.

127. K. Diehl, C. Quick, S. Matthias-Maser, S. K. Mitra, and R. Jaenicke, "The ice nucleating ability of pollen Part I: laboratory studies in deposition and condensation freezing modes," Atmospheric Research, vol. 58, no. 2, pp. 75–87, 2001. · ·

128. K. Diehl, S. Matthias-Maser, R. Jaenicke, and S. K. Mitra, "The ice nucleating ability of pollen: Part II. Laboratory studies in immersion and contact freezing modes," Atmospheric Research, vol. 61, no. 2, pp. 125–133, 2002. · ·

129. P. J. DeMott, K. Sassen, M. R. Poellot et al., "African dust aerosols as atmospheric ice nuclei,"Geophysical Research Letters, vol. 30, no. 14, 2003.

130. P. Seifert, A. Ansmann, S. Groß et al., "Ice formation in ash-influenced clouds after the eruption of the Eyjafjallajökull volcano in April 2010," Journal of Geophysical Research, vol. 116, no. D20, 2011. · ·

131. Y.-S. Choi, R. S. Lindzen, C.-H. Ho, and J. Kim, "Space observations of cold-cloud phase change,"Proceedings of the National Academy of Sciences of the United States of America, vol. 107, no. 25, pp. 11211–11216, 2010. · ·

132. P. W. Boyd and M. J. Ellwood, "The biogeochemical cycle of iron in the ocean," Nature Geoscience, vol. 3, no. 10, pp. 675–682, 2010. · ·

133. L. Bopp, K. E. Kohfeld, C. Le Quéré, and O. Aumont, "Dust impact on marine biota and atmospheric CO_2 during glacial periods," Paleoceanography, vol. 18, no. 2, 2003. · ·

134. R. Röthlisberger, M. Bigler, E. W. Wolff, F. Joos, E. Monnin, and M. A. Hutterli, "Ice core evidence for the extent of past atmospheric CO_2 change due to iron fertilisation," Geophysical Research Letters, vol. 31, no. 16, 2004. · ·

135. F. Lambert, B. Delmonte, J. R. Petit et al., "Dust—climate couplings over the past 800,000 years from the EPICA Dome C ice core," Nature, vol. 452, no. 7187, pp. 616–619, 2008. · ·

136. J. K. B. Bishop, R. E. Davis, and J. T. Sherman, "Robotic observations of dust storm enhancement of carbon biomass in the North Pacific," Science, vol. 298, no. 5594, pp. 817–821, 2002. · ·

137. P. W. Boyd, C. S. Wong, J. Merrill et al., "Atmospheric iron supply and enhanced vertical carbon flux in the NE subarctic Pacific: is there a connection?" Global Biogeochemical Cycles, vol. 12, no. 3, pp. 429–441, 1998.

138. A. Martínez-Garcia, A. Rosell-Melé, W. Geibert et al., "Links between iron supply, marine productivity, sea surface temperature, and CO_2 over the last 1.1 Ma," Paleoceanography, vol. 24, no. 1, 2009. · ·

139. B. A. Maher, J. M. Prospero, D. Mackie, D. Gaiero, P. P. Hesse, and Y. Balkanski, "Global connections between aeolian dust, climate and ocean biogeochemistry at the present day and at the last glacial maximum," Earth-Science Reviews, vol. 99, no. 1-2, pp. 61–97, 2010. · ·

140. L. J. Hoffmann, E. Breitbarth, M. V. Ardelan et al., "Influence of trace metal release from volcanic ash on growth of Thalassiosira pseudonana and Emiliania huxleyi," Marine Chemistry, vol. 132-133, pp. 28–33, 2012. · ·

141. E. P. Achterberg, C. M. Moore, A. Henson et al., "Natural iron fertilization by the Eyjafjallajökull volcanic eruption," Geophysical Research Letters, vol. 40, no. 5, pp. 921–926, 2013. ·

142. M. T. Jones and S. R. Gislason, "Rapid releases of metal salts and nutrients following the deposition of volcanic ash into aqueous environments," Geochimica et Cosmochimica Acta, vol. 72, no. 15, pp. 3661–3680, 2008. · ·

143. R. C. Hamme, P. W. Webley, W. R. Crawford et al., "Volcanic ash fuels anomalous plankton bloom in subarctic northeast Pacific," Geophysical Research Letters, vol. 37, no. 19, Article ID L19604, 2010. · ·

144. D. Lockwood, P. D. Quay, M. T. Kavanaugh, L. W. Juranek, and R. A. Feely, "High-resolution estimates of net community production and air-sea CO_2 flux in the northeast Pacific," Global Biogeochemical Cycles, vol. 26, no. 4, 2012. ·

145. A. Lindenthal, B. Langmann, J. Paetsch, I. Lorkorwski, and M. Hort, "The ocean response to volcanic iron fertilization after the eruption of Kasatochi volcano: a regional biogeochemical model study,"Biogeosciences, vol. 10, pp. 3715–3729, 2013.

146. T. R. Parsons and F. A. Whitney, "Did volcanic ash from Mt. Kasatochi in 2008 contribute to a phenomenal increase in Fraser River sockeye salmon (Oncohynchus nerka) in 2010?" Fishery Oceanography, vol. 21, pp. 374–377, 2012.

147. S. McKinnell, "Challenges for the Kasatochi volcano hypothesis as the cause of a large return of sockeye salmon (Oncorhynchus nerka) to the Fraser River in 2010," Fisheries Oceanography, 2013. ·

148. B. R. Jicha, D. W. Scholl, and D. K. Rea, "Circum-Pacific arc flare-ups and global cooling near the Eocene-Oligocene boundary," Geology, vol. 37, no. 4, article 303, 2009. · ·

149. S. M. Cather, N. W. Dunbar, F. W. McDowell, W. C. McIntosh, and P. A. Scholle, "Climate forcing by iron fertilization from repeated ignimbrite eruptions: the icehouse-silicic large igneous province (SLIP) hypothesis," Geosphere, vol. 5, no. 3, pp. 315–324, 2009. · ·

150. R. C. Bay, N. Bramall, and P. B. Price, "Bipolar correlation of volcanism with millennial climate change,"Proceedings of the National Academy of Sciences of the United States of America, vol. 101, no. 17, pp. 6341–6345, 2004. · ·

151. R. C. Bay, N. E. Bramall, P. B. Price et al., "Globally synchronous ice core volcanic tracers and abrupt cooling during the last glacial period," Journal of Geophysical Research, vol. 111, no. D11, 2006. · ·

152. S. Bains, R. D. Norris, R. M. Corfield, and K. L. Faul, "Termination of global warmth at the Palaeocene/Eocene boundary through productivity feedback," Nature, vol. 407, no. 6801, pp. 171–174, 2000. · ·

153. P. Censi, L. A. Randazzo, P. Zuddas, F. Saiano, P. Aricò, and S. Andò, "Trace element behaviour in seawater during Etna›s pyroclastic activity in 2001: concurrent effects of nutrients and formation of alteration minerals," Journal of Volcanology and Geothermal Research, vol. 193, no. 1-2, pp. 106–116, 2010. · ·

154. J. L. Sarmiento, "Atmospheric CO_2 stalled," Nature, vol. 365, no. 6448, pp. 697–698, 1993.

155. A. J. Watson, "Volcanic iron, CO_2, ocean productivity and climate," Nature, vol. 385, no. 6617, pp. 587–588, 1997.

156. L. M. Mercado, N. Bellouin, S. Sitch et al., "Impact of changes in diffuse radiation on the global land carbon sink," Nature, vol. 458, no. 7241, pp. 1014–1017, 2009. · ·

157. R. A. Scasso, H. Corbella, and P. Tiberi, "Sedimentological analysis of the tephra from the 12–15 August 1991 eruption of Hudson volcano," Bulletin of Volcanology, vol. 56, no. 2, pp. 121–132, 1994. · ·

158. S. L. De Silva and G. A. Zielinski, "Global influence of the AD 1600 eruption of Huaynaputina, Peru,"Nature, vol. 393, no. 6684, pp. 455–458, 1998.

159. C. MacFarling Meure, D. Etheridge, C. Trudinger et al., "Law dome CO_2, CH_4 and NCO_2O ice core records extended to 2000 years BP," Geophysical Research Letters, vol. 33, no. 14, 2006. · ·

160. C. D. O›Dowd, M. C. Facchini, F. Cavalli et al., "Biogenically driven organic contribution to marine aerosol," Nature, vol. 431, no. 7009, pp. 676–680, 2004. · ·

161. P. Liss, A. Chuck, D. Bakker, and S. Turner, "Ocean fertilization with iron: effects on climate and air quality," Tellus B, vol. 57, no. 3, pp. 269–271, 2005. · ·

162. C. Bonadonna and A. Foch, "Ash Dispersal Forecast and Civil Aviation Workshop—Consensual Document," 2011, https://vhub.org/resources/503.

Emerging Trend in Natural Resource Utilization for Bioremediation of Oil – Based Drilling Wastes in Nigeria

Iheoma M. Adekunle[1], Augustine O. O. Igbuku[2], Oke Oguns[3], and Philip D. Shekwolo[2]

[1]Environmental Remediation Research Group, Department of Chemical Sciences (Chemistry), Federal University Otuoke, Bayelsa State, Nigeria

[2]Restoration of Ogoniland Project Team, Shell Petroleum Development Company, Port Harcourt, Nigeria

[3]Remediation Team, Shell Petroleum Development Company, Port Harcourt, Nigeria

INTRODUCTION

Background

Nigeria is a country endowed with diverse mineral and natural resources among which is petroleum, a pivot to the national economy and sustainable development. In the past five decades, petroleum exploration and production activities have brought national economic boom but not without some aches. Acts of sabotage such as crude oil theft, pipeline bunkering and artisanal refining added to accidental spills and operational failures all combine to aggravate the oil-related aches. Oil spill into the environment, stemming from either acts of sabotage or operational failures, ultimately lead to environmental pollution with petroleum hydrocarbons [1, 2]. Petroleum mining or drilling is another factor to petroleum hydrocarbons in the environment. Most of the adverse impacts of oil spill/ petroleum hydrocarbons in the environment are experienced in the oil bearing communities, located in the Niger Delta region of the country; prominent among them being the Ogoni land pollution incidence reported by United Nations Environment Programme [1]. Petroleum exploration and production activities are strongly associated with drilling operations for oil mining. Accordingly, the extraction of petroleum resources from the earth is achieved by drilling activities. A developed drilling concept, irrespective of technological advancement, has its technical challenges, process requirements and environmental issues [3]. Drilling fluids, also referred to as drilling muds are used to enhance drilling activities via suspension of cuttings, pressure control, stabilization of exposed rocks, provision of buoyancy, cooling and lubricating.

Types of Drilling Fluids (muds): There are basically two categories of drilling fluids namely (i) aqueous drilling muds or water based muds (WBMs), which consist of fresh or salt water containing a weighting agent, usually barite ($BaSO_4$), clay or organic polymers and various inorganic salts, inert solids, and organic additives to modify the physical properties of the mud so that it functions optimally and (ii) non-aqueous drilling fluids (NADFs), which comprise all non-water dispersible base fluids such as oil based muds (OBMs) and synthetic based muds (SBMs) [2]. Comparative evaluation of oil based muds and water based muds shows that OBMs offer advantages over WBMs for the reasons that [3]:

- OBMs are more suitable to drill sensitive shells, allowing drilling faster than the WBMs, providing excellent shale stability
- they are more adequate to drill formulations where bottom hole temperatures exceed WBMs tolerance, especially in the presence of contaminants such as water, gases, cement, salt and temperature up to 550F
- OBMs resist formation salt leach out
- they are characterized by thin filter cakes and the friction between the pipe and wellbore is minimized, thus, reducing the risk of differential sticking and are especially suited for highly deviated and horizontal wells
- the drill of low pore pressure formations is easily accomplished, since mud weight can be maintained at a weight less than that of water (as low as 7.5 ppg)
- corrosion of pipe is controlled since oil is the external phase and coats the pipe. The oils are non-conductors and the additives are thermally stable, hence, do not form corrosive products
- bacteria do not thrive long in OBMs
- there is the possibility of using OBMs over and over again and can be stored over long periods of time since bacterial growth is suppressed
- OBM packer fluids are designed to be stable over long periods of time even when exposed to high temperature and provide long-term stable packers since additives are extremely temperature stable. Properly designed, such packer fluids can suspend weighting materials over long periods of times.

In other words, regarding shale stability, penetration rate, high temperatures, drilling salts, lubrication, low pore pressure formations, corrosion control, re-use and packer fluids, OBMs offer advantages over WBMs. It is therefore, obvious that though WBMs are more environmentally benign, they are only satisfactory for less demanding drilling of conventional vertical wells at medium depths, whereas OBMs are more suited for greater depths or in directional or horizontal drillings, which exert greater stress on drilling apparatus. As a result, OBMs are more frequently used in petroleum industries for drilling purposes. The composition of OBMs include: petroleum base fluid, weighting agent and other chemical additives.

Drill Cuttings: During drilling, particles of crushed rocks produced by the grinding action of the drill bit as it penetrates the earth are referred to as drill cuttings (DC). DCs are, therefore, a mixture of rocks and particulates released from geological formulations in the drill holes made for crude oil drilling and are usually coated with the drilling fluid. Consequently, DCs are largely influenced by the chemical composition of drilling muds [2, 4].

The resultant spent OBM and drill cuttings (drilling wastes) consist of hydrocarbons, water, soils, heavy metals and water soluble salts such as chlorides and sulphates [3, 4]. Drilling wastes, which are toxic due to the presence of hydrocarbons, heavy metals and other chemical additives, if not properly treated before disposal, pose serious environmental hazards and risk to public health. Sequel to these, best practices in the management of drilling wastes cannot be over emphasized.

Health and Environmental Effects Associated With Drilling Wastes

Health effects linked to drilling wastes are traceable to the basic components such as the drilling fluid and additives:

Health Effects Associated with Drilling Fluids: These health effects are attributed to the physical and chemical properties of the drilling fluids. In oil based drilling wastes, the base oil stem from petroleum stream such as crude oil, diesel (gasoil) and kerosene, which cause skin irritation. Consequently, the most commonly observed health effect associated with drilling fluids is skin irritation. Other effects include headache, nausea, eye irritation and coughing. Routes of exposure in human are dermal, inhalation, oral and some other miscellaneous routes. On exposure to drilling fluid, petroleum hydrocarbons tend to remove natural fat from the skin, which results in skin drying and cracking. These conditions allow compounds to permeate through the skin leading to irritation and dermatitis. Susceptibility to these health effects varies with individual resistance capacity and conditions of poor personal/environmental hygiene. High aromatic content fluids, especially diesel fuel contain significant levels of carcinogenic polynuclear aromatic hydrocarbons (PAHs). Diesel fuels may also be genotoxic due to high proportions of 3-7 ring PAH [2]. Skin-painting studies in mice showed that, irrespective of the level of PAH, long-term dermal exposure to diesel fuels can cause skin tumours, an effect attributed to chronic skin irritation. In humans, chronic irritation may cause small areas of the skin to thicken, eventually forming rough wart-like growths, which may become malignant. Health effects from chronic exposure to PAHs may include cataracts, kidney damage, liver damage and jaundice. Naphthalene, a specific PAH, can cause the breakdown of red blood cells, if inhaled or ingested in large amounts. Animals exposed to levels of some PAHs over long periods in laboratory studies, developed lung cancer from inhalation and stomach cancer from ingesting PAHs in food [2].

Other hydrocarbon constituents of drilling fluids are the mono-aromatics popularly referred to as BTEX (benzene, toluene, ethylbenzene and xylene). BTEX compounds are very volatile, hence, will readily evaporate in warm/hot climates of tropical regions, resulting in higher concentrations in the vapor phase. As a result, there is the possibility of exposure to human via inhalation. Exposure to high concentrations of these hydrocarbons via inhalation may result in hydrocarbon induced neurotoxicity, a non-specific effect resulting in headache, nausea, dizziness, fatigue, lack of coordination, problems with attention and memory, gait disturbances and narcosis [2].

Health Effects Associated with Additives: In addition to the irritancy of the drilling fluid hydrocarbon constituents, several drilling fluid additives may also have irritant, corrosive or sensitizing properties. Various additives include emulsion stabilizers, pH adjusters, wetting agents, viscosifiers and fluid-loss reducing agents. For instance, calcium chloride ($CaCl_2$) has irritant properties and emulsifiers (such as polyamine) have been associated with sensitizing properties [3]. Specific chemical additives vary with locations.

Environmental Effects Associated with Drilling Wastes

Apart from health effects, environmental hazards associated with drilling wastes include land, water and air pollution [5]:

- *Land Pollution*: Farming is the major land use system in Nigeria, especially in the Niger Delta region [1]. The most significant in this aspect of environmental pollution in Nigeria is thus farmland pollution. Consequences include alteration in soil physical, biological and chemical properties, loss of soil fertility, stunted plant growth and reduced crop productivity. These lead to reduced food security and compromised food safety.

- *Aquatic Pollution:* Large percentage of the oil spill gets spread over the surface of the aquatic system resulting in anaerobic environment in the water, below the surface. This leads to death of the natural flora and fauna where oxygen is the key element for their respiration; adversely affecting fishing profession [1]

- *Air Pollution:* volatile organics such as benzene, toluene, ethylbenzene and xylene could have elevated concentrations in the air, leading to atmospheric pollution and consequent adverse environmental and health impacts.

Oil well drilling processes generate large volumes of drill cuttings and spent mud in the country. Drilling wastes, therefore, add to hazardous petroleum waste materials released in the environments of the Niger Delta region of the country [1, 6] and the management of drilling wastes is quite tasking. An environmentally friendly technique for the management of drilling wastes is necessary in all offshore and onshore operations; from seismic surveys, drilling operations, field development and production to decommissioning. The physical and chemical properties of the drilling wastes influence their hazardous characteristics and environmental impact abilities, which in turn depend primarily on: (i) nature of impacted material, (ii) concentration of pollutant /amount of waste material after release (iii) recipient biotic community and (iv) exposure duration. Exposure that causes an immediate effect is called acute exposure while long-term exposure is called chronic exposure. Either acute or chronic exposure has negative impacts.

Contemporary Treatment of Drilling Waste Materials

Worldwide, contemporary drilling waste management options include re-use, offshore discharge, re-injection and onshore treatment and/or disposal [7]. Each treatment and or disposal option has its pros and cons as highlighted in the options (thermal technologies and bioremediation techniques) discussed.

Thermal Treatment

As the name suggests, thermal technologies involve the use of high temperatures to reclaim hydrocarbon contaminated materials [8]. Thermal treatment is mostly used in treating organic compounds. Additional treatment may be necessary for metals and salts depending on the final fate of the wastes. Thermal treatment technologies are designed for a fixed land based installation; however, a few mobile units exist. Two commonly practiced thermal treatment technologies are thermal desorption and incineration methods.

Thermal Desorption Method

Thermal desorption is an environmental remediation process that uses heat to increase the volatility of contaminants by the use of a series of equipment (desorber and oxidizer) such that the hydrocarbons and water are separated or removed from the solid matrix. It is normally carried out between the temperature range of 250-650°C. At these temperatures both the lighter and heavier hydrocarbons are removed and collected or thermally oxidized by further heating to a temperature of over 850°C. The resulting solid residue has essentially no residual hydrocarbons (having been oxidized), but does concentrate salts and heavy metals. Depending upon the success of process used, recovered hydrocarbons can be used as fuel or re-used as base fluid in the drilling fluid system and the resulting solid can be disposed of in a landfill or may be used in construction (of roads and bricks). Economical, operational and environmental implications of thermal desorption include:

- Effective removal and recovery of hydrocarbons from solids
- Possibility of recovering base fluid and end - product could be used for brick making
- Low potential for future liability
- Requires short time
- High cost of handling environmental issues

- Large volume of wastes is required to justify the cost of operation
- Requires tightly controlled process parameters
- High operating temperatures can lead to safety risks
- Requires several operators
- Heavy metals and salts are concentrated in residual solids
- Process water contains some emulsified oil
- Residue ash requires further treatment before disposal
- End product is sterile and can no longer support plant Life.

Incineration Method

Incineration involves (i) heating oil based mud and drill cuttings to a higher temperature range (1200-1500°C) in direct contact with combustion gases and (ii) oxidizing the hydrocarbons [8]. Solid/ash and vapor phases are generated. The gases produced from this operation may be passed through an oxidizer, wet scrubber, and bag house before being vented to the atmosphere. Stabilization of residual materials may be required prior to disposal to prevent constituents from leaching into the environment. Incineration of drilling wastes occurs in rotary kilns, which incinerate any waste regardless of size and composition. Incineration systems are designed to destroy only organic components of waste; however, most drilling wastes are non-exclusive in their content and therefore will contain both combustible organics and non-combustible inorganic materials. By destroying the organic fraction and converting it to carbon (IV) oxide and water vapor, incineration reduces waste volume. Inorganic components of wastes fed to an incinerator cannot be destroyed, only oxidized. The major inorganic materials are chemically classified as metals. Generally, these metals will exit the combustion process as oxides of the metals that enter. Economical, operational and environmental implications of incineration are as listed:

- Low potential for future liability
- High cost per volume
- Heat produced could be used for energy generation
- High energy cost
- Requires air pollution control equipment because of safety concerns

- At high temperatures, salts can form acid components
- Air emissions pose environmental concerns.

   In line with best practices, for thermal technologies, there is need for proper placement of end product. Demonstration of sufficient compliance with current regulations and adequate safety measures to cater for the potential risks of exposure to high temperatures.

Bioremediation Technique

Bioremediation technique relies on the ability of microorganisms (mostly combination of bacteria) to feed on the hydrocarbons (HCs) as substrate, converting them into carbon dioxide, water and harmless clean solids; and the ability of some of the HCs to biodegrade over time. But in most cases, the native microorganisms are often overwhelmed by the extent of the hydrocarbon contamination and thus would require external nutrients to boost (bio-stimulation) their activity and ability to take up the HCs at a faster rate. In other cases, the native microorganisms may be needing help from their kind or other species of micro-organism which are grown or inoculated (bio-augmentation) in the laboratory and then introduced in the habitat of the native micro-organisms. Bioremediation could be carried out at the site of contamination (in-situ bioremediation technique) or off the site of contamination (ex-situ bioremediation technique). Bioremediation technologies include land farming, use of bioreactors, biopiles and compost- based technologies. Economical, operational and environmental implications of conventional bioremediation technique [9, 10, 11, 12, 13, and 14] include:

- Relatively inexpensive
- Requires simple equipments and eliminates transportation cost as drill wastes could be treated on site
- Less capital but may be labour-intensive.
- Low maintenance cost; being a simple technology process that requires few machines, there are few delays due to equipment down-time
- Process is fairly flexible and can be used for most drill wastes including OBM, NADFs, previously extracted materials and newly drilled cuttings

- Proven technology
- Requires a considerable period of time to complete a process
- Appropriate bacteria and nutrient selection could be a daunting task
- In cases where bacteria are inoculated and brought on site, adaptability to their new environment may hamper their performance
- Minimal operation hazards
- Environmentally friendly: once the contaminants have been degraded, the microbial population reduces considerably as they have used up their food source
- Less impact on the environment as residue from process (TPH < 1%) may require no further treatment and could be used for agricultural purposes.

Recommended best practices for bioremediation technology include ensuring (i) proper initial physical, biological and chemical characterizations to determine extent of organic and inorganic contamination, (ii) required skill and persistence for the selection of several combinations of bacteria and nutrients that can provide the desired result (iii) proper periodic tillage to provide for proper aeration that facilitates degradation of the HCs and (iv) an accurate and appropriate TPH level check in between treatment process in order to monitor progress of the remediation process. Choice of waste management options typically considers local regulations, environmental assessment, cost/benefit analysis and the composition of the drilling wastes. The Department of Petroleum Resources [15] via the Environmental Guidelines and Standards for the Petroleum Industry in Nigeria (EGSPIN) stipulated guidelines on drill cuttings discharge for inland / near-shore and offshore deep water in order to minimize the adverse impact on the surrounding environment. These requirements call for an appropriate drill cuttings treatment prior to disposal in order to meet the stipulated conditions.

Review of Emerging Trend in the Treatment of Drilling Waste Materials in Nigeria

There are scientific evidences showing that drilling wastes generated in the country contain toxicants that are of environmental concerns. For instance, the reports of [16] on the determination of selected physical and chemical parameters including metals concentrations in a certain drill cutting dump site in the country. Results from their study showed that oil and grease on the surface and 20 feet around the waste dump area were above the specified limit [15]. There was also lack of plant growth noticed in the study, attributed to depletion of nitrogen, phosphorus and potassium values below threshold levels for plant growth. The reports of [4] on hydrocarbon and some metal contents of drilling muds and cuttings generated during the drilling of Igbokoda onshore oil wells gave total petroleum hydrocarbon (TPH), aliphatic hydrocarbon (AH) and polycyclic aromatic hydrocarbon (PAH) as generally exceeding stipulated limits by both national and international agencies. The studies of [17] on the compositional distribution and sources of polynuclear aromatic hydrocarbons (PAHs) in Nigerian oil-based drill-cuttings, showed that the total initial PAHs concentration of the drill cuttings was 223.52 mg/kg while the initial individual PAHs concentrations ranged from 1.67 to 70.7 mg/kg, dry weight, with a 90% predominance of the combustion-specific 3-ring PAHs.

The commonly employed remediation techniques for drilling wastes in Nigeria appear to be thermal technologies. However, due to economical, operational and environmental implications of these thermal technologies; search for more acceptable techniques commenced. There is scarcity of literature on the use of natural resource materials for the remediation of drilling wastes in Nigeria. The few literature resources showed that a large percentage is still at the bench-scale platform. For instance, [18] isolated *Staphylococcus sp.* from oil-contaminated soil that was treated with 1% drilling fluid base oil (HDF-2000). Their study revealed that *Staphylococcus sp.,* is a strong primary utilizer of the base oil and has potential for application in bioremediation processes involving oil-based drilling fluids. On the other hand, the effectiveness of 2 bacterial isolates (*Bacillus subtilis* and *Pseudomonas aeruginosa*) in the restoration of oil-field drill-cuttings contaminated with polynuclear aromatic hydrocarbons was studied by [19]. In that study, a mixture of 4 kg of drill cuttings and

0.67 kg of top-soil were fed into triplicate plastic reactors labeled A1 to A3, B1 to B3, C1 to C3 and O1 to O3. These were left quiescent for 7 days under ambient conditions, followed by the addition of 20 mL working solution of pure cultures of Bacillus sp and Pseudomonas sp (each of cell density 7.6×10^{11} cfu/mL) to reactors A1 - A3 and B1 - B3 respectively. Another 20 mL working solution containing both cultures at cell density 1.5×10^{12} cfu/mL was added to reactors C1 - C3. The working solution was added to each reactor (excluding the controls, O1 - O3) every 2 weeks. Mixing and watering of the set-ups were carried out at 3 days interval under ambient temperature of 30°C for a period of 6 weeks. Results showed that the predominant 3-ring PAHs, which made up 90% w/w of the total PAHs concentration of 223.52 mg/kg, were degraded below detection and the 4-ring PAHs were reduced from 4 to 0.6% by Pseudomonas while Bacillus reduced 3 and 4-ring PAHs respectively to 0.2 and 0.8%. Their works revealed that Pseudomonas degraded 3 and 4-ring PAHs relatively better than Bacillus. Both strains of bacteria degraded 5 and 6-ring PAHs below detection limits. Furthermore within the 3-ring PAHs, each of the strains of bacteria reduced phenanthrene to approximately 0.2%, whereas both degraded homologues acenaphthylene, acenaphthene and fluorene as well as anthracene below detection limits. For 4-ring PAHs, Pseudomonas degraded fluoranthene and benzo[a]anthracene. Bacillus also degraded benzo[a]anthracene below detection limits. Pseudomonas was able to reduce pyrene and chrysene to 0.3 and 0.2% respectively; whereas Bacillus reduced fluoranthene, pyrene and chrysene to 0.1, 0.01 and 0.4% respectively. However, treatment with the mixed culture resulted in limited degradation of 5-ring PAHs particularly in the fourth week, which was attributed to the phenomena of co-metabolism and inhibition.

The works of [20] compared the potentials of bio-augmentation and conventional composting as bioremediation technologies for the removal of PAHs from oil-field drill-cuttings. From a mud-pit, close to a just-completed crude-oil well in the Niger Delta region of Nigeria, 4000 g of drill cuttings was obtained and homogenized with 667 g of top-soil (to serve as microbes carrier) in three separate reactors (A, B and C). The bio-augmentation of indigenous bacteria in the mix was done by adding to reactors A and B a 20-mL working solution (containing 7.6×10^{11} cfu/mL) of pure culture of Bacillusand Pseudomonas, respectively, while a 20-mL working solution (containing 1.5×10^{12} cfu/

mL) of the mixed culture of *Bacillus* and *Pseudomonas* was added to reactor C. The bio-preparation was added to each reactor (excluding the control) every two weeks for six weeks. The composting experiment was conducted in a 10-litre reactor in which 4000 g of drill cuttings, 920 g of topsoil and 154 g of farmyard manure and poultry droppings were homogenized. Mixing and watering of the set-ups were carried out at 3 days interval under ambient temperature over a period of six weeks. Results showed that initial individual PAHs concentrations in the drill cuttings ranged from 1.67 to 70.7 mg/kg dry weight, with a predominance of combustion-specific 3-ring PAHs (representing 90% of a total initial PAHs. After the bioremediation exercise that lasted for 42 days, total PAHs in the drill cuttings were reduced from 223.52 to 4.25 mg/kg, representing a 98.1% reduction. Away from the use of microbial strains in the treatment of drilling wastes, a bench-scale investigation was carried out by [21] to demonstrate the efficacy of technique referred to as 'Dispersion by Chemical Reaction (DCR) technology".This particular method involved the use of hydrophobized calcium oxide (CaO) to form a dry, soil-like material that could be useful in construction works.

On the other hand, after the study on the response of four phytoplankton species in some sections of Nigeria coastal waters to crude oil in controlled ecosystem [22], that revealed the adverse impacts; a multidisciplinary environmental remediation research group (ERRG) was inaugurated with the mandate to embark on innovative, cutting-edge research and development (R & D) initiative, aimed at the development of an indigenous technology for an eco-friendly technique in the treatment of soils, sediments, sludge and drilling wastes polluted by petroleum hydrocarbons, using natural products of Nigeria origin. The goal of ERRG is to translate the technology from bench-scale to field scale and come out with on- the - shelf products that will find use for both onshore and offshore remediation works. The first phase of the R & D initiative was the exploration of the remediation potential of conventional composting technology based on the results from the works of [23]. A good start was the production of a scientifically formulated and classified compost bulk [24] that are potentially viable for environmental remediation projects [25] and able to biodegrade petroleum hydrocarbons embedded in soil and related matrices [26]. The next phase was to assess public acceptance of the principles of this technology, which culminated to the reports of [27] on population perception impact on value-added solid waste disposal in developing countries, a case study of Port Harcourt City. The feedstock utilized in product formulations in this emerging, indigenous and innovative technology is 100% biodegradable and very abundant in the Nigerian environment. Consequently, the technology has been categorized by stakeholders [27] as:

- eco-friendly environmental remediation technique
- waste to wealth initiative
- waste to resource initiative
- value-added waste management option
- a contribution to the promotion of local material development that has the potential for:
- wealth creation
- job creation
- poverty alleviation
- sound environmental management of hydrocarbon polluted wastes from the petroleum industries.

ERRG observed that either conventional composting technology or bioremediation via utilization of pure microbial isolates/ strains has limitations in terms of serving the practical needs of the petroleum industry in Nigeria with regards to meeting (i) regulatory remediation targets at close – out of project and (ii) project delivery time. Subsequently, through series of bench-scale and screen house remediation investigations, products were formulated to enhance the speed of bioremediation process using nano-scale green catalysts, a technique that matured into Compost - based Nanotechnology in Bioremediation (CNB-Tech). The research group then subjected the CNB-Tech products to different scientific evaluations in order to ascertain (i) efficiency on biodegradation of petroleum hydrocarbons in oily wastes such as crude oil impacted soils, sludge and drilling wastes (drill cuttings and oil-based mud) and (ii) environmental impacts with emphasis on soil quality. Published works on assessment and prognosis of products' impact on soil quality include:

- Assessing the effect of bioremediation agent from local resource materials in Nigeria on soil pH [28]

- Impact of bioremediation formulation from Nigeria local resource materials on moisture contents for soils contaminated with petroleum [29]
- Assessing and forecasting the impact of bioremediation product derived from Nigeria local raw materials on electrical conductivity of soils contaminated with petroleum products [30]
- Soil temperature dynamics during bioremediation of petroleum products using remediation agent from Nigerian local resource materials [31].

Other works on CNB-Tech products' evaluations including (i) effect on soil heavy metal dynamics and (ii) impact on soil microbial species population and diversity are being considered elsewhere for publication. Having recorded a huge success during the laboratory scale investigations where maximum of 4000g of sample bulk and freshly hydrocarbon contaminated soils (similar to the quantities used by other investigators) [19, 20] were treated, it became necessary to assess the efficiency of CNB-Tech products on waste materials with complex nature and higher degree of hydrocarbon pollution. This aspiration was realized in collaboration with the Remediation Department of Shell Petroleum Development Company (SPDC), Port Harcourt, Nigeria through the University Liaison Team of SPDC. Sequel to this, pilot-scale projects were commissioned to evaluate the efficiency of CNB-Tech products on the degradation of hydrocarbon compounds in the following petroleum impacted materials:

- Hydrocarbon polluted clay soils from Ejama-Ebubu legacy site of SPDC
- Hydrocarbon polluted carbonized soil from Ejama-Ebubu legacy site of SPDC
- Hydrocarbon polluted sludge from Ejama-Ebubu legacy site of SPDC
- Oil-based mud and drill cuttings generated from SPDC operations.

Ejama Ebubu is one of SPDC's legacy sites of up to 42 year long pollution as at the time of study in 2011 [1]. In this chapter, the efficacy of CNB-Tech products in the biodegradation of petroleum hydrocarbons in oil-based drilling wastes (OBM-DC) is presented.

Research Justification

The treatment of drilling wastes, especially OBM-DC in an environmentally sound manner is a challenging task due to the complex nature of the wastes. The most popular technique adopted for the treatment of OBM-DC, thermal desorption [15] has its accompanying environmental concerns. For instance, thermal treatment technologies are associated with prohibitive capital and operational cost implications, threatening environmental consequences in addition to high occupational hazards and generation of secondary waste stream that has to be treated at extra high cost before final disposal. Consequently, there is need for a pragmatic shift to seek alternative techniques that will address the need of the oil and gas sector in the management of drilling wastes in terms of remediation target delivery time and compliance to regulatory standards in Nigeria. Regulatory standards for close-out of remediation projects vary from one country to another and success factors of a given technology are dependent on indices such as:

- climatic conditions
- geographical characteristics of the location
- nature and complexity of contamination
- expected utility of the end-products of the remediation exercise

It then becomes evident that a successful remediation technology in one part of the globe may not necessarily be efficient in another region, pointing to the need to look inward for a more practical approach to solving the environmental challenges posed by petroleum hydrocarbon polluted waste streams in Nigeria [1]. Having run laboratory, bench-scale and screen-house remediation works using CNB-Tech products on fresh hydrocarbon contaminated soils, it became necessary to conduct pilot scale remediation works on more challenging waste streams such as weathered petroleum impacted soils, sludge, sediment, oil- based drilling mud and drill cuttings, hence this project.

Research Objectives

The current study comprised three major objectives:

- to conduct a review on the emerging trends in the treatment and related studies for drilling wastes in Nigeria,

- to assess the efficiency of an indigenous and innovative application of compost - based nanotechnology in bioremediation (CNB-Tech) in biodegradation of hydrocarbons found in oil-based mud and drill cuttings; generated by a petroleum industry in Nigeria
- to investigate the beneficial utility of the remediation end-product for agricultural purpose (crop production), which is a major land use system in Nigeria.

RESEARCH METHODOLOGY

The research methodologies employed in this study were:

- Literature review to provide an insight to the current and emerging trend in the treatment of drilling waste materials in the country and
- Practical, ex-situ, pilot scale execution of biodegradation of hydrocarbon compounds in oil-based mud and drill cuttings generated by an oil company in Nigeria using an indigenous and innovative biotechnological (CNB-Tech) approach anchored on the use of natural resource materials of Nigeria origin.

Pilot-Scale Remediation of Oil-Based MUD and Cuttings Using CNB-Tech Method

This study was carried out during the 2010/2011 Sabbatical Programme of the University Liaison Team of Shell Petroleum Development Company (SPDC); in conjunction with the Remediation Department of SPDC, Port-Harcourt, Nigeria. The indigenous remediation products (CNB-Tech products) prepared from cellulosic natural resource materials and biogenic nanopolymers of Nigeria origin used for this pilot remediation study, were denoted as (i) Ecorem, (ii) Bioprimer and (iii) Biozator. The last two products are solids that are transformed to the aqueous form before use while the first product is used in the solid form.

Project Site Description

The present pilot-scale project, for the purposes of adequate monitoring and efficient execution, was carried out in the Industrial Area of Shell Petroleum Development Company, Port Harcourt, Rivers State; known as "Shell IA". The earmarked project area was a relatively isolated open green field within Shell IA and according to design, a temporary sheltered facility constructed to suit the project design was erected at the site and all necessary health and safety issues were taken into consideration. The sheltered project facility comprised of three major units:

- Remediation execution section: where actual remediation took place
- Phyto-analytical section: where effects on plant life were investigated
- Mini- chemical laboratory: where necessary onsite chemical evaluations were conducted.

Pilot Scale Remediation Procedure

The batch of oil-based mud and drill cuttings (OBM-DC) used in this study was generated from SPDC's operations and supplied by one of the company's certified vendors. During the conveyance procedure for OBM-DC, chain of custody document and waste stream tracking manifest was observed. Basic highlights for CNB-Tech application mode are outlined in Figure 1. Pretreatment involved recovery of free phase base fluid and stabilization involved modification of viscosity parameter.

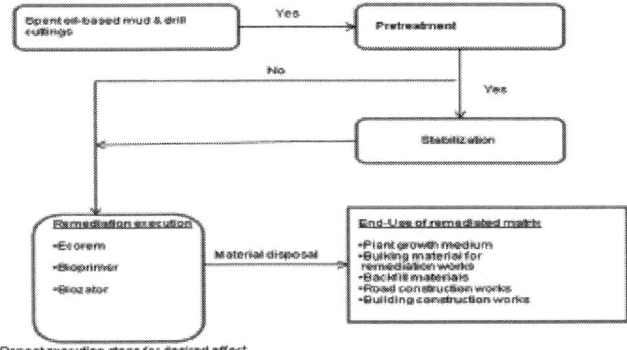

Figure 1: Application model of CNB-Tech remediation method.

The biocell utilized for the remediation execution was designed by the research group, locally fabricated and lined with appropriate PVC materials. The procedures involved in the pilot remediation exercise are described as follows: A biocell of total dimension 15 m^3 was sub-divided to smaller units of 3 m x 1 m x 1 m to allow for five times replication. Ecorem (a CNB-Tech product) was placed in the cells prior to loading of oil-based drilling mud and cutting (OBM-DC) that have been previously conditioned using intervention CNB-Tech products. As the initial microbial population in OBM-DC was less than 2.0 x 10^3 cfu/mL, Ecorem was introduced at 10% by weight of waste materials. Using mechanical means, OBM-DC and Ecorem were homogenized and allowed to incubate for about 12 to 24 hours in order to trigger and stimulate natural microbial activities. CNB-Tech products (Bioprimer and Biozator) were then applied to saturate the contents in the biocells, which was followed by homogenization using mechanical devices. A CNB-Tech product was added to the leachate (process fluid) to immobilize inorganic constituents (especially metals) before recycling the leachate into the treatment network in such a manner that no leachate was produced as a by-product for discharge into the environment. OBM-DC that received no treatment served as control. Both controls and test units were subjected to the same environmental conditions.

System Maintenance and Monitoring: During remediation, the system was monitored for relevant environmental factors such as moisture content (I), pH (II), nitrogen content (III) and temperature (IV) using standard procedures of gravimetry for I, probe method via a calibrated pH meter for II, Kjedahl method for III and calibrated mercury in glass thermometer for IV. These environmental factors were maintained at the required range. Remediation lasted for 33 days: 6 days for actual treatment and 27 days for material fallow and recovery periods during which the treated materials were conditioned with a CNB-Tech product (Ecorem) for use as plant growth medium.

In order to validate the efficacy of this technology, representative composites were sent to an International Laboratory (RespirTeK Consulting Laboratory and affiliate Laboratories based in the United States of America) for physical, chemical and microbial assessments. RespirTek is ISO/EC accredited and certified. Three other laboratories that are based in Nigeria (certified by national regulatory bodies) were also involved in sample collection and analyses. Laboratories that participated in this study were:

- Technology Partners International Nigeria Limited, Port Harcourt - Nigeria
- Laser Engineering and Resources Consultants Limited, Port Harcourt- Nigeria
- Fugro Nigeria Limited, Port Harcourt, Nigeria
- RespirTek Consulting Laboratory - United States of America

Sample Collection

At the end of the pilot remediation project using CNB-Tech products, treated materials were moved from the biocells and spread out on PVC impermeable membranes (each of dimension 650 cm for length and 248 cm for width), homogenized using mechanical means and air-dried with occasional homogenization of samples. The dry samples were returned into the biocells where further homogenization procedure was carried out. Sampling containers were sent by RespirTEK Consulting Laboratory, USA for their own use.

General Sample Collection: Using mechanical means, treated and dried samples in the cells were thoroughly homogenized for one week. In order to collect sample from a particular replicate, each replicate was subdivided into 4 equal parts; representative fractions were collected from the different parts and recombined to give a composite sample of 1kg.

BTEX Sampling: Standard sampling kit for BTEX, sent by RespirTEK Consulting Laboratory, was utilized for the purpose. In this procedure, homogenized samples were collected from the cells using "Terra Core" sampling device. Using a 40 mL glass VOA vial containing appropriate preservatives and with the plunger seated in the handle, the Terra Core was pushed into freshly homogenized sample until the sample chamber was filled to the capacity of 5g. All sample particulates (debris) were removed from the outside of the Terra Core sampler and the sample plug was pushed into the mouth of the sampler. Excess soil that extended beyond the mouth of the sampler was removed. The plunger was then seated in the handle and rotated until it aligned with the slots in the body. The mouth of the sampler was placed into the 40 mL VOA vial containing the preservatives and sample extruded by pushing the plunger down. The lid was quickly placed back on the 40 mL VOA vial. It was ensured that when capping the 40 mL VOA vial, sample debris was removed from the top of the vial.

All samples were appropriately labeled and recorded in the chain of custody form before shipping to the USA laboratory by courier. Two Laboratories in Nigeria also collected samples for analyses, following standard procedures. The third laboratory in Nigeria was only involved in the analysis of materials using infrared and UV-absorption spectroscopic methods.

Physicochemical Analysis and Microbial Assessment

Statement from quality control and quality assurance unit (QA/QC) of RespirTek Laboratory, USA showed that all analyses were conducted following procedures set forth by the ISO/IEC 17025:2005 accreditation program standards for which the laboratory holds certification. Quality assurance systems and quality control criteria were strictly followed. The following parameters were determined:

- Total petroleum hydrocarbons (TPH)

- Monoaromatic hydrocarbons: benzene, toluene, ethylbenzene and xylene (BTEX). For xylene, ortho -, meta - and para- derivatives were assessed

- PAHs: a total of 17 PAH compounds: (i) naphthalene, (ii) acenaphthylene, (iii) acenaphthene, (iv) fluorene, (v) phenanthrene, (vi) anthracene, (vii) fluoranthene, (viii) pyrene, (ix) benzo (a) pyrene, (x) chrysene, (xi) benzo (b) fluoranthene, (xii) benzo (k)fluoranthene, (xiii) benzo (a) pyrene, (xiv) dibenz(a,b) anthracene, (xv) benzo (ghi)perylene, (xvi) 2-methylnaphthalene and (xvii) indeno (1,2,3-cd) pyrene

- Metals: barium (Ba), calcium (Ca), copper (Cu), lead (Pb), mercury (Hg), Nickel (Ni), Sodium (Na), Potassium (K), cadmium (Cd), zinc (Zn) and arsenic (As), a metalloid

- Miscellaneous parameters: pH, salinity, nitrogen, phosphorus, total organic carbon and electrical conductivity.

- Microbial activity: assessment of 48 hr and 96 hr microbial activities of both remediation end-product and contaminated material (control) was conducted by the USA based laboratory. Total hydrocarbon utilizing bacteria as well as total microbial count were assessed by the Nigerian based laboratories.

Hydrocarbon compounds were analyzed using Gas chromatographic method, microbial assessment was carried out using heterotrophic plate count method and metals were determined using atomic absorption spectroscopic technique. All the other parameters were carried out using standard procedures such as described in [24, 25, 32]. The CNB-Tech products (Bioprimer and (Biozator) were characterized using infrared and UV-visible spectroscopic methods. The basic characteristics of Ecorem have already been reported in [24, 25] but was slightly enhanced, in this study, for case specificity.

Assessment of Seed Germination Potential of Treated Samples

The remediated materials used in this evaluation were not mixed with external soil and no external fertilizer material was added to the remediated soil. Seed germination potential (SGP) of treated samples were assessed and only viable maize seedlings were used for this purpose. In a remediated material matrix (4kg material contained in an experimental plastic pot), 6 seedlings of maize were sown. This was replicated three times. All together, 18 (6 x 3) seedlings were used to evaluate this effect. Similar set- ups were also established for the untreated oil – based mud and cuttings, which served as control systems. This gave a total of 18 (6 x 3) seeds tested for germination potential for the test systems and 18 seedlings for the control media. This phase of the evaluation lasted for 7 days.

Assessment of Process Fluid (Leachate) Effect on Plant Growth

Adequate leachate (process fluid) management strategy was put in place as leachate generated during remediation was recycled into the remediation process. However, this evaluation was to ensure or to prove that in the event of any leachate seepage there would be reduced environmental risk. This phytotoxicity assessment was carried out using a cereal (corn: Zea mays L.,) as an indicator crop and indices of toxicity were (i) root length and (ii) plant height. Experimental systems constituted of the following set-ups, where FS is dilution factor and SF stands for farm soil:

- Farm soil + tap water (Code: FS + water). This served as control system for (ii) and (iii)
- Farm soil + stock leachate (Code: FS + LDF-0). This served as control system for (iii)
- Farm soil + diluted leachate series:
  - Farm soil + leachate DF-1 (Code: FS + LDF-1)
  - Farm soil + leachate DF-2 (Code: FS + LDF-2)
  - Farm soil + leachate DF-3 (Code: FS + LDF-3)
  - Farm soil + leachate DF-4 (Code: FS + LDF-4)

For this assessment, bulk farm soil sample, obtained from a village (K-dere, part of Ogoniland) in Rivers State, was used. Soil was sieved through a mesh and transferred at 1.5 kg per pot and designated pots were treated to 70% approximate field capacity (determined against gravity) using equal volume of appropriate fluid (water, stock leachate or diluted leachate). The systems were allowed to stabilize for 2 weeks after which viable maize seedlings were sown at 3 per pot. As the plants grew, the soil systems were treated with equal volumes of the appropriate fluid to maintain appropriate moisture level, as required by plant. Experiment lasted for 2 weeks, at the end of which the heights were recorded and plants harvested. Caution was exercised to ensure that roots were not destroyed during harvest. Root lengths were then recorded and mean values per pot calculated for each parameter.

Evaluation of Beneficial Utilization of End-product

Similar to the case in Section 2.4, in this evaluation, the remediated matrix was not mixed with any type of soil, neither was any external fertilizer administered. At close - out of the pilot-scale remediation project, the remediated materials were air dried, primed with one of CNB-Tech products (Ecorem) at a specified loading scheme and then utilized as a growth media. Primed end-products were transferred at 4 kg per pot of 4 liter capacity. Three indicator crops used for this project were:

- Corn (*Zea mays L.,*)
- Green leafy vegetable (Fluted pumpkin: *Telfairia Occidentalis*)
- Cassava (*Manihot esculenta Crantz*)

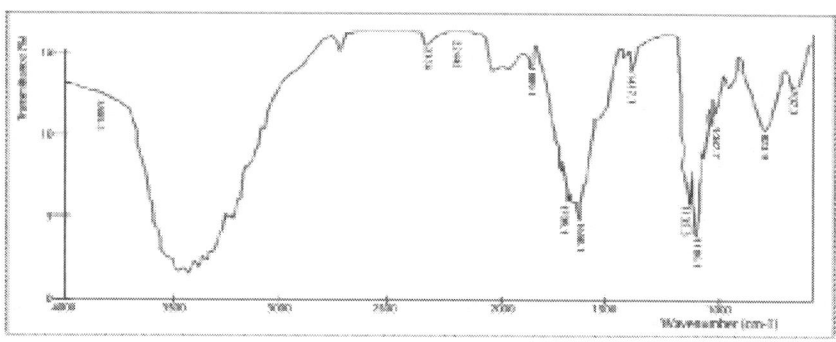

Figure 2: Infrared spectrum of Bioprimer, a CNB-Tech remediation product.

The crops were used because they are commonly grown and consumed in the Niger Delta region of the country. Due to time constraint, duration of investigation varied for the crops, the longest being up to 130 days for green leafy vegetable (Fluted pumpkin: *Telfairia Occidentalis*) while corn (*Zea mays L.,*) and cassava (*Manihot esculenta Crantz*) were grown for 2 and 3 weeks respectively. Untreated OBM-DC served as a control and farm soil served as a second control.

Statistical Analysis

Data generated in this study were subjected to statistical evaluations using SPSS software for Windows, version 17.0. Descriptive statistics were applied to evaluate mean and standard deviation. Paired sample T-Test and One-way analysis of variance (ANOVA) were applied to identify significant variations among treatments as appropriate. Pearson correlation was used to ascertain significant relationships.

RESULTS

Typical Infrared Spectra of Two CNB- Tech Remediation Products

The infrared absorption spectra of two CNB-Tech products (Bioprimer and Biozator) utilized in this pilot scale study are presented in Figures 2 and 3. Both spectra showed absorption peaks in the region of 4000 to 600 cm^{-1}.

Major information from the infrared spectra were: strong, broad absorption band of oxygen-hydrogen (O-H) of an alcohol (aryl/aliphatic) and N-H absorption bonds around 3500 - 3300 cm^{-1}; carbon-oxygen double bond (C=O) absorption band found around 1750 – 1500cm^{-1} This could be carbonyls of ester (RCOOR), aldehyde (RCHO), ketone (RCOR) and acid (RCOOH). C-N bond of nitrogenous matter falls in the end of the range; C-O bond around 1200 – 1000 cm^{-1} and of carbon-hydrogen (C-H) bond for aromatic moieties found below 1000cm^{-1} [33].

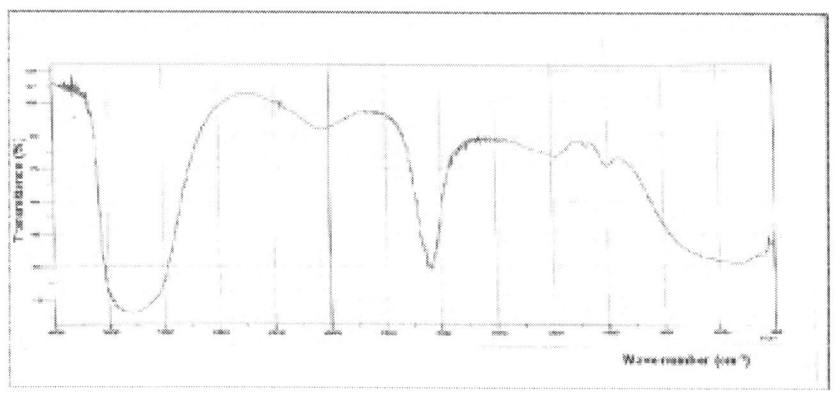

Figure 3: Infrared spectrum of Biozator, a CNB-Tech remediation product.

Initial Characteristics of the Drilling Wastes

The results presented in this paper were largely those obtained from the International laboratory. Table 1 contains the initial characteristics of the drilling wastes (oil-based mud and cuttings).

Table 1: Initial characteristics of the oil -based drilling mud and cuttings used in this pilot scale study

S/N	Parameter	Concentration
Inorganics		
1.	Arsenic (mg/kg)	6.69
*2.	Cadmium	Not determined
3.	Barium(mg/kg)	765
4.	Calcium(mg/kg)	87300
5.	Copper(mg/kg)	35.90
6.	Lead(mg/kg)	161
7.	Mercury(mg/kg)	0.036
8.	Nickel(mg/kg)	12.3
9.	Sodium(mg/kg)	493
10.	Potassium(mg/kg)	1930
11.	Zinc(mg/kg)	144

12.	TKN (%)	0.0357
13.	Phosphorus (%)	0.0291
*14.	pH	10.2
*15.	Electrical conductivity (mSm^{-1})	Not determined
16	Total organic carbon (%)	Not determined
17..	Salinity (mg/kg)	4300
BTEX compounds		
1.	Benzene	0.0198
2.	Ethylbenzene	0.827
3.	m- and p-xylene	0.532
4.	o-xylene	0.924
5.	toluene	1.910
PAH Compounds		
1.	Naphthalene(mg/kg)	1.94
2.	Acenaphthylene(mg/kg)	BDL
3.	Acenaphthene(mg/kg)	BDL
4.	Fluorene(mg/kg)	2.54
5.	Phenanthrene(mg/kg)	0.78
6.	Anthracene(mg/kg)	BDL
7.	Fluoranthene(mg/kg)	BDL
8.	Pyrene(mg/kg)	BDL
9.	Benzo (a) anthracene(mg/kg)	BDL
10.	Chrysene(mg/kg)	BDL
11.	Benzo(b)fluoranthene(mg/kg)	BDL
12.	Benzo (k)fluoranthene(mg/kg)	BDL
13.	Benzo(a)pyrene(mg/kg)	BDL
14.	Dibenz(a,h)anthracene(mg/kg)	BDL
15.	Benzo(g,h)perylene(mg/kg)	BDL
16.	2-methylnapthalene(mg/kg)	5.39

17.	Indeno(1,23-cd)pyrene(mg/kg)	BDL
	Total PAH(mg/kg)	10.65
Total petroleum hydrocarbon		
1.	TPH (mg/kg)	79 200

[i]: *Parameters not determined by the USA laboratory but quantified by Nigerian based laboratories

Results indicated the presence of inorganic constituents and organics (hydrocarbons compounds). Regarding inorganics, soft metal contents increased in the order: Na (493 mg/kg) < K (1930 mg/Kg) < Ca (87, 300 mg/kg). The elemental ratios were 177 for Ca/Na, 45 for Ca/K and 4 for K/Na. Heavy metal concentrations increased in the order: Hg < As < Ni < Zn < Cu < Pb < Ba. In terms of hydrocarbon contents, total concentrations of polynuclear aromatic hydrocarbon (PAH) compounds was 10.65 mg/kg with concentrations of the individual components (Figure 4) increasing as phenanthrene (0.78 mg/Kg: 7%) < naphthalene (1.94 mg/kg; 18%) < fluorene (2.54mg/kg; 24%) < 2-methylnapthalene (5.39 mg/kg; 51%). Results on monoaromatics (BTEX), shown in Figure 5, gave a total concentration of 4.213 mg/kg out of which toluene constituted the highest fraction (45.34%), followed by xylene (34.56%), ethylbenzene (19.63%) and benzene (0.47%). Total xylene concentration was 1.456 mg/kg out of which ortho-xylene constituted 63.46% while meta- and para-xylenes gave 36.54% of the total (1.456 mg/kg).

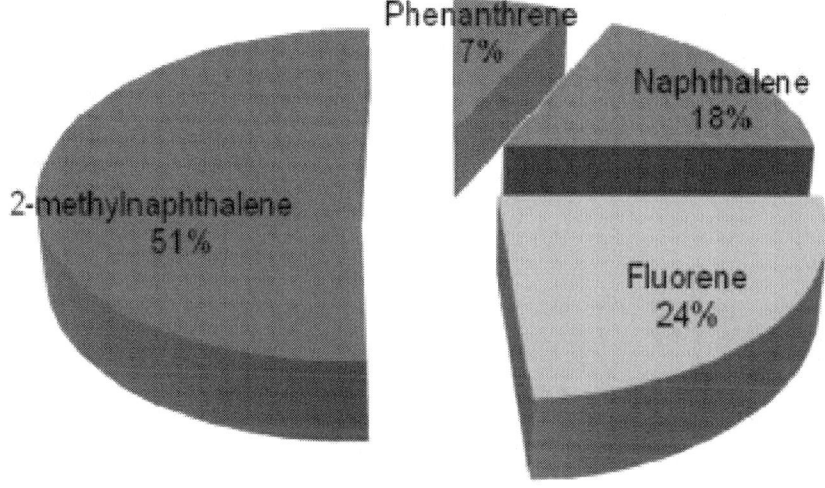

Figure 4: Percentage distribution of individual components of PAH relative to the total concentration.

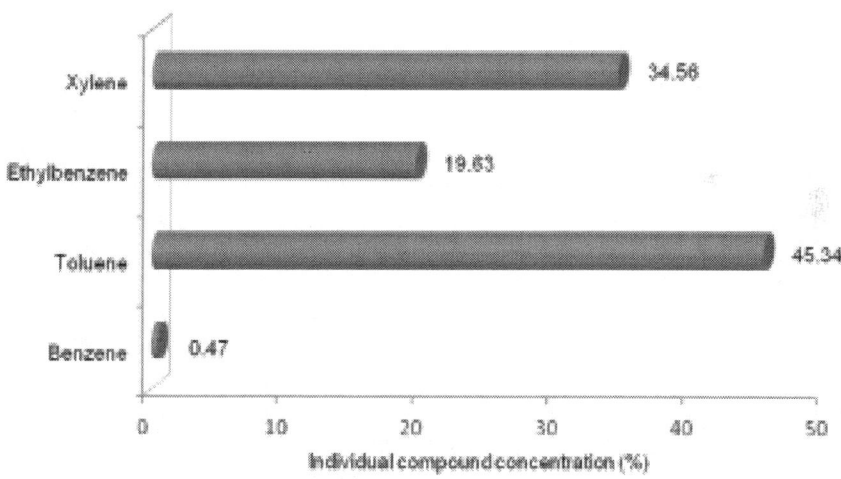

Figure 5: Percentage distribution of individual components relative to the total BTEX concentration.

Results on Petroleum Hydrocarbon Degradation

By application of CNB-Tech products, the initial TPH concentration of 79, 200 mg/kg decreased to 1888.67 ±161. 20 mg/kg. The difference in these two values was a mean TPH concentration of 77 311.33 ± 161.20 mg/kg. This difference corresponds to the total concentration of hydrocarbon compounds degraded or destroyed by the applied treatment. The initial concentration (79, 200 mg/kg) and the degraded fractions (in replicates of three) are presented in Figure 6. Specifically, results on hydrocarbon degradation (Figure 7) revealed 98% degradation for TPH, 100% degradation for BTEX and 100% degradation for PAH. Reduction in TPH level by 99% was obtained by the Nigerian laboratories.

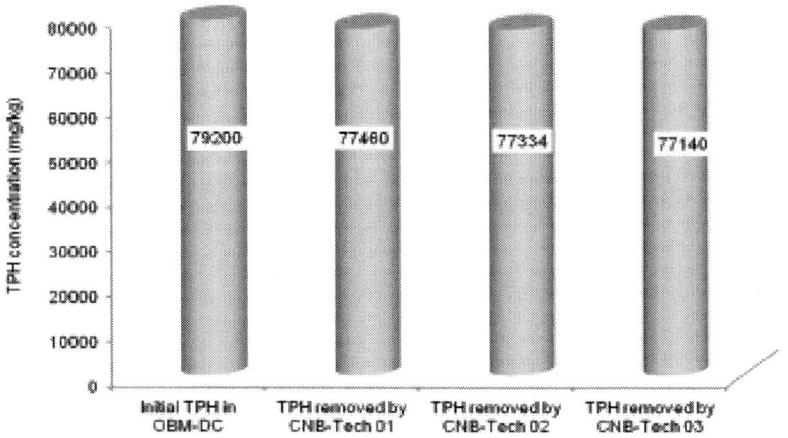

Figure 6: Graph showing concentrations of degraded TPH relative to the initial concentration.

Table 2: Qualitative results for the remediated media

S/N	Parameter	Remarks for contaminated medium	Remarks for remediated medium
1.	Appearance	Viscous, pasty and solid interfaced in oil suspension	Transformed to non-viscous, non-sticky crumby humus soil appearance
2.	Color	Light brown	Treated matrix had characteristic dark color of humus soil
3.	Odor	Presence of strong hydrocarbon odor	Complete disappearance of hydrocarbon odor in all the treated media and all treated samples exhibited clean earthy smell
4.	Sheen test	Strong oil sheen in water suspension	Complete disappearance of oil sheen in water suspension

Results on qualitative assessments of the untreated OBM-DC and remediated material in terms of appearance, odor, and color and sheen test are contained in Table 2 and Figure 8 depicts the materials' appearances before and after remediation.

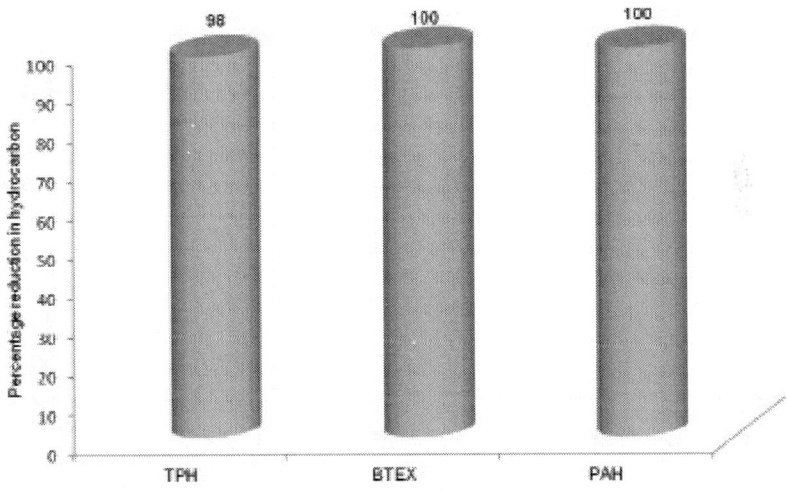

Figure 7: Percentage degradation of hydrocarbon compounds in the drilling wastes by applied CNB-Tech products.

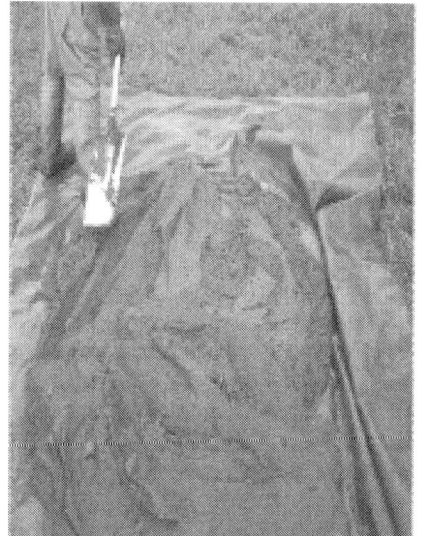

Figure 8: Photographs showing the materials before and after bioremediation by the application of CNB-Tech products.

Results on Inorganic Constituents of the CNB -Tech Treated Materials

Descriptive statistics of selected inorganic constituents found in the treated media are presented inTable 3. Changes in their concentrations relative to the initial values are presented in Figure 9. For instance, the initial pH value was reduced to 7.90 from 10.20, corresponding to 23% reduction. Likewise, the following reductions were obtained: 62% for Ca, 46% for As, 44% for Cu, 70% for Pb, 100% for Hg, 57% for Ni and 37% for Zn. The concentrations of some elements such as nitrogen, phosphorus and potassium were elevated. The nitrogen-phosphorus-potassium (NPK) status, as affected by treatment, is presented in Figure 10. Nigerian laboratories obtained the same trend for NPK status. Based on the results from USA, CNB-Tech remediation option applied in this study raised the nitrogen level from 0.036% to 0.096%, raised phosphorus level from 0.0291% to 0.312%, increased potassium by 1.4 fold (Figure 10) and sodium by 3 folds. The USA based laboratory did not analyze for total organic carbon and electrical conductivity but the Nigerian based laboratory did and recorded electrical conductivity in the range of 1956 to 2063 mSm^{-1} with a mean value of 2003 ± 54 mSm^{-1} before treatment. After remediation, the electrical conductivity of the end products ranged from 594 to 696 mSm^{-1} and a mean value of 640± 52 mSm^{-1}. From the mean values, there was a 68% reduction in electrical conductivity.

Table 3: Concentrations of some inorganic parameters in the treated materials

S/N	Element	Minimum	Maximum	Mean	Standard error	Standard deviation	Sample population
1.	pH	7.70	8.20	7.90	0.15	0.26	3
2.	Nitrogen (%)	0.070	0.130	0.096	0.016	0.028	3
3.	Phosphorus (%)	0.280	0.360	0.312	0.026	0.046	3
4.	Potassium (%)	0.50	0.77	0.61	0.08	0.14	3
5.	Copper (mg/kg)	18.10	21.70	20.10	1.06	1.83	3
6.	Zinc (mg/kg)	79.30	110	92.67	9.08	15.73	3
7.	Nickel (mg/kg)	3.99	7.05	5.29	0.92	1.59	3
8.	Calcium (mg/kg)	28900	39200	33466	3030	5248	3
9.	Arsenic (mg/kg)	2.50	4.85	3.59	0.68	1.18	3
10.	Lead (mg/kg)	5.87	54.80	27.06	14.50	25.12	3

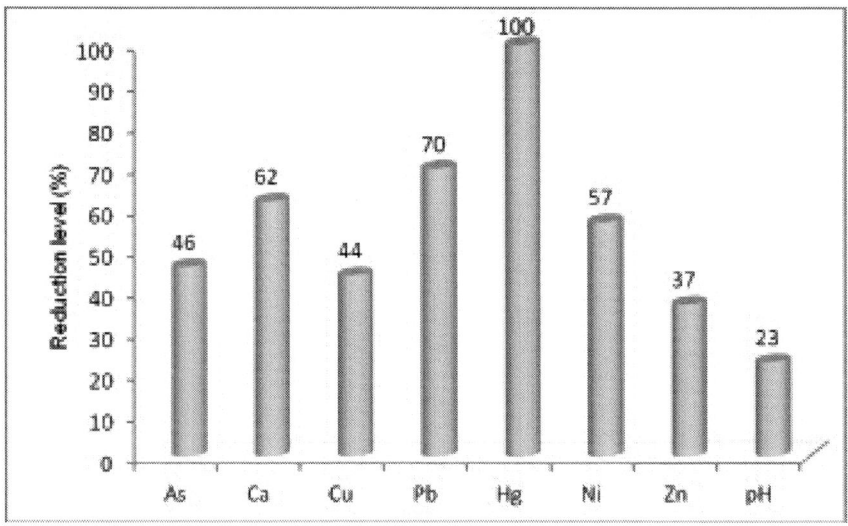

Figure 9: Reductions in some inorganic constituents of the drilling materials treated by CNB-Tech.

Total organic carbon ranged from 2.95 to 3.06% with a mean of 2.99± 0.06% before remediation and increased to 3.84 to 3.93% with a mean of 3.88 ± 0.05%; corresponding to an increase by 23%. Before remediation, Cd concentration varied from 6.70 to 7.60 mg/kg, with a mean value of 7.03± 0.49 mg/kg. After treatment, the metal concentration ranged from 0 to 1.80 mg/kg with an average of 1.05 ± 0.94 mg/kg. By the two mean values, cadmium level was reduced by 85% due to applied CNB-Tech products.

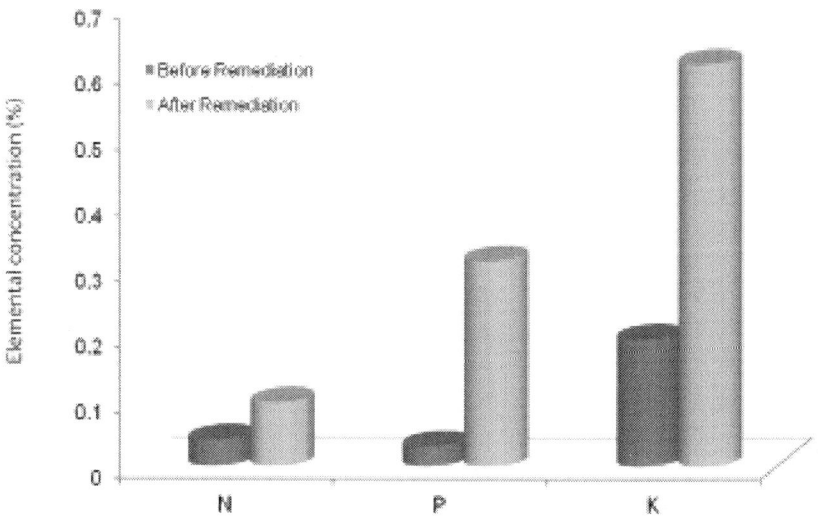

Figure 10: Nitrogen-phosphorus-potassium status before and after treatment as obtained by the USA based laboratory.

Results on Microbial Activity

The digital photographs of heterotrophic plate count results are shown in Figure 11. Microbial activities assessed on the untreated and treated samples revealed that the contaminated oil-based mud and cuttings (no. 1 in Figure 11), contained some indigenous microorganisms of up to 1.9×10^3(cfu/mL) while the CNB-Tech remediated samples recorded up to a maximum of 3.15×10^7 cfu/mL. An illustration of microbial enumeration for 48-hr and 96 hr counts are presented in Figure 12.

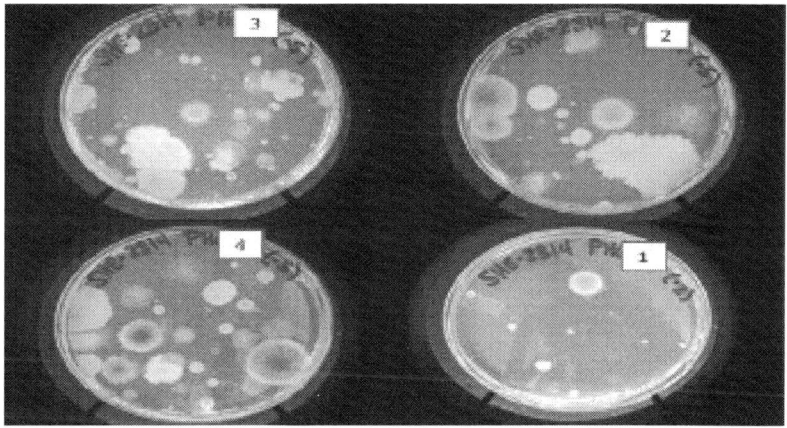

Figure 11: Heterotrophic plate count digital photographs for untreated OBM-DC (1) (before remediation) and replicates (2, 3, 4), after remediation using CNB-Tech method.

At 48 hr microbial activity assessment, maximum total microbial population of 1.9×10^3 cfu/mL was obtained for untreated OBM-DC and in the materials remediated by the application of CNB-Tech products, it was 1.45×10^7 cfu/mL. These two values were significantly different at $p \leq 0.05$. At 96 hr microbial activity assessment, a total microbial population of 2.4×10^3 cfu/mL was obtained for untreated OBM-DC and 3.15×10^7 cfu/mL for the remediated matrices. Results showed that within 48 hours, the microbial activity of the remediated matrices excelled over the untreated by over 7,000 folds and at 96 hours, it excelled by over 13, 000 folds, indicating rapid multiplication of microbial activity by CNB-Tech products which also increased with time.

Results on Phytotoxicity Assessment of Remediated Samples

Toxicity on Seed Germination Potential

The contaminated OBM-DC did not allow the germination of maize seedlings. Out of the sown 18 seedlings, none germinated. The untreated OBM-DC therefore, gave 100% toxicity to seed germination potential (SGP) of maize. On the contrary, all the 18 maize seedlings sown in the CNB-Tech remediated matrices germinated (Figure 13). Hence, resulting in 100% positive effect on SGP, indicating that the treated matrices exhibited 0% toxicity to seed germination.

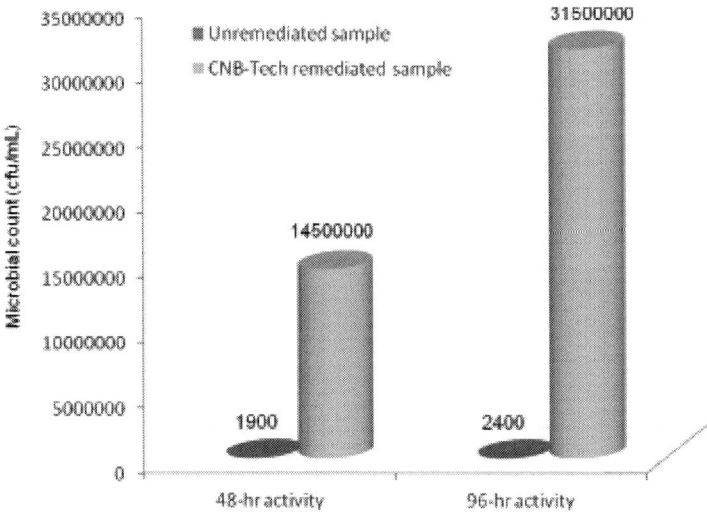

Figure 12: Microbial activity at 48 –hr and 96-hr counts for untreated oil-based drilling wastes and CNB-Tech remediated samples.

Figure 13: Germinated maize seedlings growing in treated media with picture taken on day 4 of growth.

Results on Beneficial Use of Remediation End Product

Figure 14, shows a cross-section of the treated materials (during recovery period) being aerated in preparation for use as plant growth media.

Figure 14: A cross section of project technical staff preparing the treated drilling wastes (OBM-DC) for use as plant growth media.

During the recovery phase of the remediated end-product, treated materials were allowed to lie fallow in order to establish natural processes as a sign of wellbeing and restoration. In this project, after the fallow period, early indications of material restoration were:

- spontaneous vegetative growth,
- the presence of larva within the spontaneously grown green vegetation,
- butterflies and small birds perching on the surface of the material, which could not take place before treatment

Remediated materials supported the growth of fluted pumpkin (*Telfairia occidentalis*). A cross-section of the green leafy vegetable at over 100 days of growth and that of cassava, at one week of growth, growing in the treated materials are shown in Figure 15. Narrowing to the height of *Telfairia occidentalis*, the mean height for crops grown in the untreated OBM-DC was 0 cm as there was complete inhibition to both germination and growth. The mean height for crops grown in

CNB-Tech remediated media was 217± 25 cm, a value higher than the mean height (187± 40 cm) of the vegetable crops grown in farm soil collected from the region. The difference in the two mean values was significant at p = 0.14. Correlation for the heights of the vegetables grown in the treated media and those grown in the farm soil gave a coefficient of 0.95 (p = 0.204).

Figure 15: Remediated drilling wastes as plant growth medium for Fluted pumpkin (*Telfairia occidentalis*) and cassava (*Manihot esculenta Crantz*).

Results on the Impact of Remediation Leachate on Plant Life

Comparative evaluations of control system (soil treated with water only), stock leachate system (soil treated with leachate without any form of dilution) and systems treated with serial dilutions of the leachate (soil treated with leachate diluted with water by factors 1, 2, 3 and 4) are presented in Table 4.

Table 4: Impact of leachate generated at the close-out of project on the root length and height of maize

S/N	System Code	Leachate effect of on vegetative growth relative to control (%)		Effect of serial dilution on plant using stock (undiluted leachate) as reference (%)	
		Height	Root length	Height	Root length
1.	FS+ Water (Control)	Reference	Reference	Not applicable	Not applicable
2.	FS + DF-0	-1.50	-23.45	Reference	Reference
3.	FS + DF-1	32.60	1.12	34.62	32.20
4.	FS + DF-2	45.01	16.37	42.22	50.02
5.	FS + DF-3	66.86	21.37	69.41	58.55
6.	FS + DF- 4	75.39	24.51	78.07	62.66

[i] - Negative sign stands for decrease. The other positive values stand for increase, FS = farm soil and DF = dilution factor

Pictorial and graphical representations of leachate impact on plant height and root length are presented in Figures 16 and 17. Relative to the control system (soil treated with water only), leachate diluted with water by a factor of 4 improved plant height by 75.39% and root length by 24.51%. Figures16 and 17gave all the systems at a glance, relating the control (FS + Water), system SF+LDF-0 (DF-0) and serial dilutions (DF-1 = FS+ LDF-1, DF-2 = FS+ LDF-2, DF-3 = FS+ LDF-3 and DF-4 = FS + LDF - 4) for plant height and root length. Evaluating the effect of leachate dilution relative to the stock (undiluted) leachate, a 4-fold dilution excelled over the stock by 78.0% for plant height and 62.66% for root length. The relationships between plant height or root length and dilution factors are given inFigure 18. Pearson correlations gave strong coefficients: plant height versus dilution factor, $r = 0.979$ ($p = 0.004$), root length versus dilution factor, $r = 0.932$ ($p = 0.021$) and plant height versus root length, $r = 0.972$ ($p = 0.006$). From the results, plant vegetative growth increased with increasing dilution of leachate.

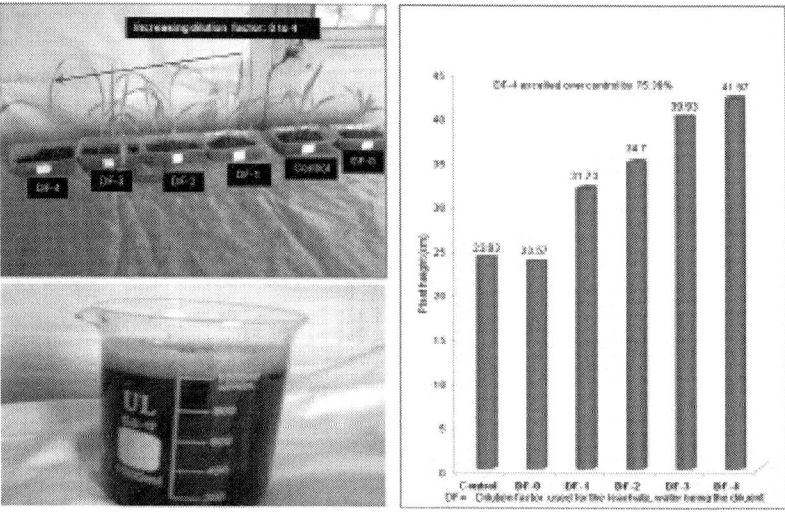

Figure 16: Pictorial and graphical representations of leachate impact on height of maize, including a picture of the stock leachate contained in a beaker.

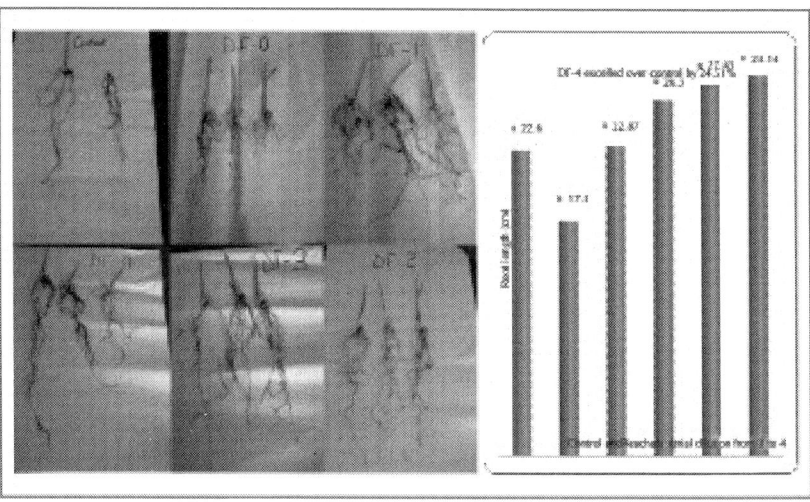

Figure 17: Pictorial and graphical representations of leachate impact on root length of maize.

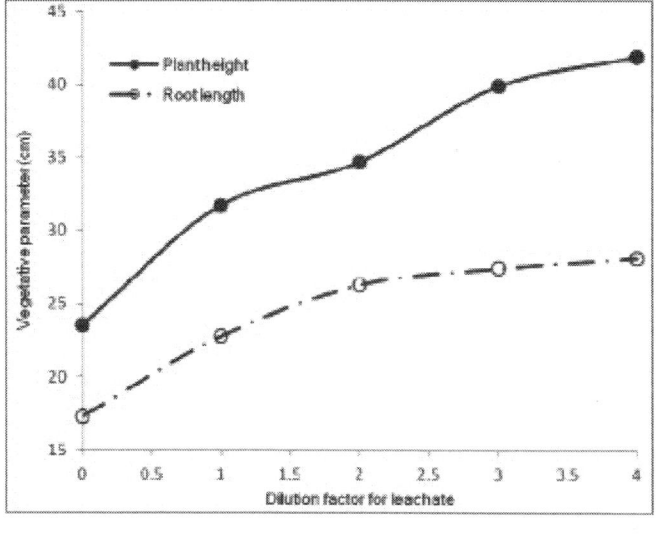

Figure 18: Relationship between plant vegetative growth and serial dilution of process fluid (leachate) generated during the remediation project.

DISCUSSION

The type of inorganic constituents and hydrocarbons found in the drilling wasting used in this study were consistent with the reports of [4, 17] but varied in concentrations. This confirms that the OBM-DC used in this study was toxic [2]. The remediation products of CNB-Tech series used in this study demonstrated a high (98 to 100%) degradation potential for the different constituents of hydrocarbon compounds found in the drilling wastes, within a short period of 6 days. This excellent performance was attributed to the chemistry, nature and operation mechanisms of the CNB-Tech formulations.

An infrared spectrum is primarily used to identify functional groups present in a molecular fragment [33]. The infrared spectra obtained for CNB-Tech products (Biozator and Bioprimer) revealed enrichment of the molecular structure of the two products with oxo- groups, indicating oxidizing functionality. The presence of C-H of aromatic nature and the O-H stretching absorption indicate the presence of both hydrophobic and hydrophilic properties, respectively, in their molecular fragments.

By implication, the remediation products are naturally endowed with:

- oxidizing ability
- polar (hydrophilic: water loving) molecular fragment
- non-polar fragment (hydrophobic: water insoluble, oil soluble) molecular fragment.

These natural endowments permit the dissolution of the products' active ingredients (solids) in water, making water the carrier medium for CNB-Tech liquid formulations. Consequently, Biozator and bioprimer are water based technical grade products. By the mentioned characteristics, the two products perform reduction and oxidation (Redox) reaction mechanisms, resulting in the degradation/ destruction of hydrocarbons compounds, without recombination to form new hydrocarbons. These absorption peaks in the infrared spectra further reveal that CNB – Tech products are natural hydrocarbon biodegradation catalysts for the following reasons:

- enhaced water solubility of hydrocarbons via sorption, hydrolysis and oxidation mechanisms
- enhanced bioavailability of hydrocarbon pollutants for microbial degradation
- increased supply of oxygen [O] molecules required for enhanced reduction –oxidation reactions in the hydrocarbon degradation process.
- surfactant property
- emulsification of hydrocarbons

The combined actions of hydrophobic molecular fragment, hydrolysis, oxidation and surfactant property of CNB-Tech products render hydrocarbons more water soluble and subsequently more available for biodegradation. Bioprimer and Biozator also emulsify hydrocarbons into droplets that can be easily assimilated by microorganisms. By these properties, the products reduce oil-water surface tension; enhance water solubility of petroleum hydrocarbons thereby enhancing the bioavailability of the contaminants (hydrocarbons) to microorganisms for both extracellular and intracellular decompositions. The two products are 100% biodegradable. The third CNB-Tech product used in this study (Ecorem: a black amorphous solid material, also 100% biodegradable) contains major and minor plant nutrient elements and via hydro-activation, naturally generates mixed consortia of microorganisms, which multiplies with time to facilitate the destruction of hydrocarbons. No engineered microorganism or externally imported microorganism was used in this study. This technology, therefore, saves time and eliminates the daunting task of isolating pure microbial strains and associated adaptability challenges linked with conventional bioremediation techniques [7, 8, 18,19, 20].

The microorganisms from Ecorem product perform the following functions:

- extracellular decomposition in which the naturally produced microorganisms secrete enzymes to breakdown large organic compounds (such as hydrocarbons) into smaller forms for easier absorption into the micro-organisms. Once the smaller compounds have been absorbed by the microorganisms, intracellular decomposition takes place

- increased microbial activity facilitated by Ecorem, results in thermophilic temperature modulations in the range of 55 to 60°C, a process that accelerates degradation of hydrocarbons, especially polynuclear aromatic aromatic hydrocarbons (PAHs). Thermophilic temperature modulations also controls thermo-sensitive pathogen to crops animals and man; killing off weeds and seeds that will be detrimental to land use of end products.

By the above described mechanisms, the CNB-Tech products were able to biodegrade petroleum hydrocarbon compounds with high efficiency (98% degradation for TPH and 100% degradation for PAHs and BTEX) within a short period of time of 6 days, relative to previous works on bioremediation. For instance, in a study of in-situ bioremediation of oily sludge via biostimultaion of indigenous microbes, conducted by [34], through the addition of manure at the Shengli oilfield in Northern China for 360 days, 58.2% reduction in TPH was achieved in test plots and 15.5% reduction in control plot. By treating 2 kg of drill cuttings with initial TPH of 806.36 mg/kg for 56 days under the conditions of composting of spent oyster mushroom (P.ostreatus) substrate, [35] recorded overall degradation of PAHs in the range of 80.25 to 92.38%. In this present study, OBM-DC used had initial TPH of 79, 200 mg/kg and was degraded by 98% within the stated short period of 6 days. In a field trial biopile composting method [36] for drilling mud polluted sites in the Southeast of Mexico with comparable TPH level of 99 300 ± 23000 mg/kg, after 180 days, TPH concentrations decreased from 99 300 ± 23000 mg/Kg to 5500 ± 700 mg/kg, corresponding to 94% degradation for amended biopile and to 22900 ±7800 mg/kg, representing 77% decrease for unamended biopile. The mean residual value of TPH (5500 ± 700 mg/kg) left in the treated matrix in their study was higher than the mean residual value (1888± 161 mg/kg) obtained in this present study.

By conducting an investigation on two bioremediation technologies (bioremediation by augmentation and conventional composting using crude manure and straw) as treatment options for oily sludge and oil polluted soil in China [12] in which the total hydrocarbon content (THC) varied from 327.7 to 371.2 g/kg (327700 to 371200 mg/kg) for dry sludge and 151.0 g/kg (151000 mg/kg) for soil for a period of 56 days; after three times of bio-preparation application, THC decreased by 46 to 53% in the oily sludge and soil. The results (98 -100% degradation) obtained from this present study was from only one dose application of CNB-Tech products. Repeated application of CNB-Tech products by two to three dose applications will achieve 100% degradation of TPH. In another instance, a 5- month field scale bioremediation of sludge matrix via the utilization of organic matter such as bark chips via conventional composting, mineral oil (equivalent to total hydrocarbons) decreased from 2400 to 700 mg/kg (70% decrease) for sludge matrix and from 700 to 200 mg/kg, corresponding to 71% decrease [14]. In treating oil sludge using composting technology in semiarid conditions for 3 months, hydrocarbons were reduced from 250 to 300g/kg (250000 to 300 000 m/kg) by 60% against reduction by 32% recorded in the control [37]. The treatment applied by [37] and consequent reduction of 60% implies that the residual hydrocarbons in the treated samples would be between 100 000 and 180 000 mg/kg unlike the results obtained in this present study that gave residual hydrocarbon of 1888.67 ±161.20 mg/kg. In a study carried out by [38], sand samples contaminated with oil spill were collected from Pensacola beach (Gulf of Mexico) and tested to isolate fungal diversity associated with beach sands and investigate the ability of isolated fungi for crude oil biodegradation. From their results, 4.7 to 7.9% biodegradation was recorded.

Elsewhere in India, Abu Dhabi and Kuwait [39], bioremediation technology was applied in field-scale degradation of hydrocarbons in different oil wastes for a period of 12 months. Table 5 illustrates different reductions in total petroleum hydrocarbons obtained in these field case studies. TPH reductions in drilling wastes were obtained in the range of 90.85 to 95.48% with residual TPH in treated samples in the range of 2600 to 10 900 mg/kg (0.26 to 1.09%).

Table 5: Reductions in TPH levels obtained in field case studies of different types of petroleum impacted wastes (soils, drill cuttings and oil-based mud) in Abu Dhabi, Kuwait and India [39]

Name of the oil Installation / type of oily waste	Quantity of oily waste (cubic meter)	Number of batches	TPH Content (%) in oily waste before and after bioremediation		% Reduction in TPH	Residual TPH in treated material (%)
			Before	After		
Abu Dhabi National Oil Company (ADNOC), Abu Dhabi / Oil contaminated drill cuttings	200	1	17.26	0.98	94.32	0.98
BG Exploration and Production India Limited (BGEPIL), India / Oil based mud (OBM)	2,428	3	5.75 – 6.23	0.26 - 0.57	95.48- 90.85	0.26 – 0.57
Bharat Petroleum Corporation Limited (BPCL), India / Oily sludge	5,000	1	19. 30 – 26.5	0.26 - 0.57	98.65- 97.85	0.26 -0.57
Cairn Energy Pty. India Limited, India / Oil contaminated drill cuttings	567	2	14.93 – 18.81	0.82 – 1.09	94.51- 94.21	1.09
Chennai Petroleum Corporation Limited (CPCL), India / Oily sludge	4,444	2	26.12	0.89	96.59	0.89
Hindustan Petroleum Corporation Limited (HPCL), India / Oily sludge	5,010	3	16.70 – 52.81	0.90 – 1.60	94.61- 96.97	0.90-1.60

Indian Oil Corporation Limited (IOCL) Refineries in India / Oily sludge (acidic + non acidic)	75,412	48	9.6 – 38.4	0.37 – 0.95	96.15-97.53	0.37-0.95
Kuwait Oil Company (KOC), Kuwait / Oil contaminated soil	778	1	4.6 – 12.75	0.09 – 0.10	98.04-99.21	0.09-0.10
Mangalore Refinery and Petrochemicals Limited (MRPL), India. / Oily sludge	2,222	2	8.35 – 19.86	0.84 – 0.97	89.84-95.12	0.84-0.97
Oil and Natural Gas Corporation Limited (ONGC) installations in India / Oily sludge & oil contaminated soil	95,499	145	12.0 – 51.5	0.5 – 1.2	95.83-97.67	0.50--1.20
Oil India Limited (OIL) , Assam / Oily sludge & oil contaminated soil	15,921	14	21.6 – 37.7	0.49 – 0.53	97.73-98.59	0.49-0.53
Reliance Energy Limited (RIL), India / Oily sludge	611	2	19.15	0.5	97.39	0.50

The residual TPH level (1888.67 ± 161.20 mg/kg) obtained in this present study was below the Environmental Guidelines and standards for the Petroleum Industry in Nigeria (EGASPIN) intervention value for mineral oil (petroleum hydrocarbon) of 5000 mg/kg [15]. By repeated application of CNB-Tech products, it is possible to meet a very strict regulatory standard for residual TPH level of less than 50 mg/kg. The changes in metal concentrations found in this study were attributed to (i) immobilization via chelate formation (ii) preferential supplementation of trace plant nutrient elements using the three products, (iii) natural electrochemical process whereby the positively or negatively charged organic molecules (generated during the natural transformation process occurring when the products were in use) bond with their counterparts in organic matter. These processes include oxidation, methylation, hydroxylation, carboxylation, coupling and polymerization [40] thereby enhancing bioavailability of the metals to microorganisms that utilize the organic matter supplied by the CNB –Tech products as energy source.

Microbial population found in a typical tropical soil under Nigerian climate is in the neighborhood of 8.19×10^6 cfu/mL [41]. Relative to this value, the population found in the contaminated OBM-DC (1.9 to 2.4×10^3 cfu/mL) showed suppressed microbial population, attributed to strong hydrocarbon (TPH level of 79, 200mg/kg) pollution. This is in agreement with the reports of [3]. The microbial population (1.45 to 3.15×10^7 cfu/mL) found in treated samples revealed restoration of soil microbial population using CNB-Tech products. It excelled over the value recorded in polluted material by over 7000 folds and higher than the value reported by [34], where TPH degraders and PAH degraders increased by one to two orders of magnitude via the addition of manure. Furthermore, the use of CNB-Tech products modified the pH value of the drilling wastes, transforming it from strongly alkaline (pH of 10) medium to pH of 7.90 medium; comparable to the 7.3 ± 0.1 obtained by [34] for bioremediated soils. The very high pH of the untreated drilling waste materials could be attributed to some of the additives in the drilling fluid. Drilling fluids contain an internal phase of brine such as calcium salts [3]. This was confirmed by the high content of Ca (87 300 mg/kg) obtained in this study for the untreated material. One dose application of CNB-Tech products reduced this concentration by up to 62%, repeated dose application would definitely bring Ca level to any desired value.

Observations made during the recovery /fallow period were signs of drastic positive change in toxicity conditions, implying reduced toxicity. Reduction of soil toxicity by bioremediation, evidenced by increase in EC50 of the soil was reported by [34]. In this study, bioremediation using CNB-Tech products reduced toxicity in treated materials relative to untreated OBM-DC, evidenced by 100% positive effect on seedling germination potential and improved crop vegetative growth. Reduced material toxicity also explains the increased microbial activity of the treated matrices in comparison to the untreated drilling wastes, obtained in this study. The agricultural potential for the remediation end-products was also manifested by:

- increased microbial activities
- increased nitrogen-phosphorus-potassium (NPK) status
- increased soil crumby nature as against very viscous and pasty characteristics of untreated drilling wastes.

These nutrient elements (NPK) enhance microbial growth, microbial population, microbial activity and consequently increase soil fertility [41]. By these, CNB-Tech products could overcome the extreme phytotoxicity [100% toxicity to seedling germination potential of maize and 100% inhibition to vegetative growth for three different types of plant (maize, fluted pumpkin and cassava)], caused by the untreated drilling waste. CNB-Tech products transformed oil-based drilling mud/cuttings to arable soil; capable of supporting seed germination and plant growth; excelling the performance of a control (farm soil apparently not impacted by drilling waste or crude oil) by 14%.

Electrical conductivity, a measure of dissolved ions in solution, is influenced by several soil physical and chemical properties such as salinity, saturation percentage, water content, bulk density, organic matter content, temperature and cation exchange capacity of the soil matrix. Impact of these influencing factors must be reflected in interpreting electrical conductivity effect on plant growth. Generally, elevated electrical conductivity and high salinity levels in agricultural soils may result in reduced plant growth and productivity or in extreme cases, the elimination of crops and native vegetation [42]. The reduction of electrical conductivity by 68% is a positive development because it demonstrates that the products could also modify the salinity of the material. In situations of very high initial electrical conductivity, there is a step-down CNB-Tech product as was carried out in this study and in situations of very low electrical conductivity, there is also a step-up CNB-Tech product as reported in a previous publication [30]. Results in this present study on excellent growth of crops planted in the remediated matrices were indicators of acceptable soil salinity level for plant growth. The beneficial use of the end-products obtained in this study for crop production were attributed to postulations based on findings from this study and previous works on this subject matter, which include:

- stimulation of beneficial microorganisms in soil, which enhances soil fertility [25]
- possible increased photosynthetic rate in plants evidenced by increased photosynthetic pigments (chlorophylls a and b) [40]
- increase in soil buffering capacity [28]
- increased soil moisture retention capacity by reducing hydrophobicity tendency [29]
- positive soil temperature modifications that enhance soil nutrient bioavailability to plants [31, 40]
- formation of stable chelates with toxic metals such as Pb, Cu and Cd in order to reduce their bioavailability to plants [40]
- preferential exclusion of the chelated toxic metals from soil solution, allowing the plant nutrient elements to be assimilated into plant cells
- improvement of soil physicochemical properties via:
- increased aeration and water retention [29]
- activation of the macro and micro nutrients in soil in forms readily assimilated by plants [30, 40]
- improvement of plant root development and growth
- improvement of seed sprout of plants and subsequent shoot growth
- improved plant biomass production [26]
- enhanced soil nitrogen, phosphorus and potassium status for improved soil fertility
- acting as plant growth hormone, having positive stimulant action for plant growth [25, 26]
- improvement of soil permeability, promoting plant drought resistance [29]
- promotion of increased soil porosity and organic matter content, hence greatly promoting the microorganism activity and improving soil fertility.

Regarding leachate generation and management during the remediation exercise; fluid (leachate) produced as remediation progressed was recycled by incorporation into the biocell and used to regulate moisture content, thereby reducing water usage and conserving water resources. Expertise applied during the project ensured that at remediation project close-out, no isolated fluid system was actually produced. Nonetheless, the assessment of leachate effect on plant growth carried out in this work was to establish the fact that even in the event of accidental release of some fluid into the environment, there would be minimal risk to the receptor biotic community. More evaluations are still ongoing in this regard. Results from this study revealed that the leachate generated, though a concentrate, supported plant growth and when diluted with ordinary tap water gave a better support; reasons being that:

- toxic petroleum hydrocarbons in the contaminated drilling wastes have been destroyed to an acceptable level, evidenced by natural foamability of the concentrated leachate. Foamability would hardly occur if oil was still present

- leachate is also enriched with plant nutrients such as nitrogen, phosphorus and potassium

The process fluid, therefore, had some fertilizer value. The percentage decreases (1.50% and 23.45%) obtained for plant height and root length respectively, for the stock leachate was attributed to concentrated level of nutrients, confirmed by better performance of dilute leachate series. Naturally, in any formulated fertilizer, plant nutrients are applied at specified concentrations otherwise may hinder plant growth. Comparative evaluations of control system (soil treated with water only), stock leachate system (soil treated with leachate without any form of dilution) and systems with serial dilutions of the leachate (soil treated with leachate diluted with water by factors 1, 2, 3 and 4) revealed that the leachates were not toxic to receptor plants. The implication of this is that in the event of occasional spill of the leachate to the adjacent environment; dilution with water is, therefore, an adequate safety measure.

The ability of the end products to sustain the growth of green leafy vegetable: fluted pumpkin (*Telfairia ocidentis*) and root tuber crop, cassava (*Manihot esculenta Crantz*) and cereal crop (maize) is a demonstration of the utility of the remediation end product. It therefore stands that the use of CNB-Tech products as a biotechnological tool for hydrocarbon degradation in drilling waste converts these waste materials into non-toxic and potentially useful end products. In addition to the beneficial use of the remediation end-product for agricultural purposes, other possible utility options, shown in Figure 19, include:

- material for road construction
- material for building construction
- substrate for the production CNB-Tech bioremediation agents
- excellent organic fertilizer for subsistence and commercial agriculture
- feedstock for bioremediation projects

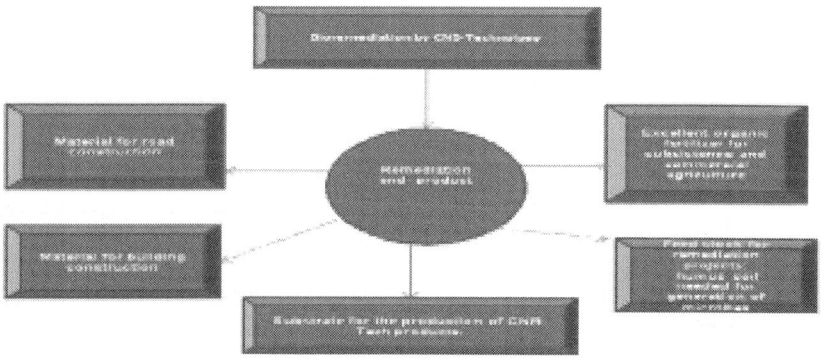

Figure 19: Potential utility of end - products from bioremediation using CNB-Tech products.

Table 6 is a comparative evaluation of economic, operational and environmental implications of thermal technologies as reported by [3] and CNB-Technology based on the results and learning from this study.

Table 6: Comparative evaluation between thermal technology and CNB-technology

S/N	Thermal Technology	CNB-Tech
1.	Effective removal and recovery of hydrocarbons from solids	Effective removal of hydrocarbons from solid
2.	Possibility of recovering base fluid and end - product could be used for brick making	Effective recovery of free phase oil and end product has other uses apart from brick making
3.	Low potential for future liability	No future liability
4.	Requires short time	Time is relatively short

5.	High cost of handling environmental issues, since end- product dispersion would be below organic layer where vegetation growth is desired	Very minimized environmental issues
6.	Large volume of wastes is required to justify the cost of operation	Cost-effective for either small or large volume of wastes
7.	Requires tightly controlled process parameters	Does not require tightly controlled process parameters
8.	Heavy metals and salts are concentrated in processed solids	Reduces heavy metals and salts concentrations in process solid
9.	High operating temperatures can lead to safety risks	Low operating temperature. Operates at ambient temperature; modulation does not exceed 60°C.
10.	Requires several operators	Does not require several operators
11.	Process water contains some emulsified oils	Process water does not contain some emulsified oils
12.	Residue ash requires further treatment	No residue ash. End-product is clean soil
13.	End product is sterile and can no longer support plant Life	End product is fertile and can support microbial and plant Life

CONCLUSIONS AND RECOMMENDATIONS

This study revealed that it is possible to harness natural, biodegradable and local resource materials of Nigeria origin; translate them to scientifically formulated products that can be utilized for efficient biodegradation of hydrocarbon polluted matrices such as oil-based mud and drill cuttings within a reasonable short period of 6 days. This technology thus converts hydrocarbon polluted oil-based mud and drill cuttings to beneficial end-products of high order reuse such as soil amendment, without the generation of secondary waste materials. Field-scale trial adopting CNB-Technology is recommended.

ACKNOWLEDGEMENTS

This project was carried out under full financial support of the Remediation Department, Shell Petroleum Development Company (SPDC), Port Harcourt, Nigeria through the University Liaison Team of the company. The support of the Oil well Team of SPDC that facilitated the procurement of oil- based mud and drill cuttings is also acknowledged.

REFERENCES

1.   United Nations Environmental Programme (UNEP), 2011. Environmental Assessment of Ogoniland. P.1-262. ISBN:978-92-807-9 Available on line at: http://postconflict.unep.ch/publications/OEA /UNEP_OEA.pdf

2.   Department of Health, Government of South Australia (DHGSA). Public Health Fact Sheet on Polycyclic Aromatic Hydrocarbons (PAHs): Health effects 2009 http://www.dh.sa.gov.au/pehs/PDF-files/ph-factsheet-PAHs-health.pdf

3.   Neff, M.M and Duxbury, MA. Composition, environmental fates, and biological effects of water based drilling muds and cuttings discharged to the marine environment: A Synthesis and Annotated Bibliography. Prepared for Petroleum Environmental Research Forum (PERF) and American Petroleum Institute. 2005. http://perf.org/pdf/APIPERFreport.pdf

4.   Gbadebo, A.M., Taiwo, A.M. and U. Eghele, U Environmental impacts of drilling mud and cutting wastes from the Igbokoda onshore oil wells, Southwestern Nigeria. Indian Journal of Science and Technology, 2010; 3(5), 504 -510.

5.   Environmental Protection Agency (EPA). An Assessment of the Environmental Implications of Oil and Gas Production: A Regional Case Study, 2008

6.   Osuji, L.C., Erondu, E.S and Ogali, R.E Upstream petroleum degradation of mangroves and intertidal shores: The Niger Delta Experience. Chemistry and Biodiversity, 2010: 7, 116 -128.

7.   Knez, D., Jerzy, A, G and Czekaj Trends in the drilling waste management. Acts Montanistica Rocnlk, 2006:11, 80-83.

8. Morillon, A., Vidalie, J.F., hamzah, U.S., Suripno and Hadinota, E.K "Drilling and Waste management", SPE 73931, Intenationa; Conference on Health, Safety and Environment in oil and gas exploration and production, 2002: March 20-22

9. Zimmerman, P.K. and Rober, J.D Oil-based drill cuttings treated by landfarming. Oil and Gas J, 1991: 12, 81-84

10. Rojas-Avelizapa, N.G., Roldan-carrillo, T., Zegarra-Martinez, H., Munez-Colunga, A.M and Fernandez-Linares A field trial for an ext-situ bioremediation of a drilling mud-polluted site. Chemosphere 2007: 66, 1595-1600.

11. Frydda, S and Randle, J.B Case study: Biological treatement of Geothermal drilling cuttings. Proceedings World Geothermal Congress, Bali, Indonesia, 25-29, 2010: 1-3.

12. Ouyang, W., Liu, H., Murygina, V., Yu, Y., Xiu, Z and Kalyuzhnyi, S Comparison of bio-augmentation and composting for remediation of oily sludge: A field-scale study in China. Process Biochemistry, 2005: 40, 3763 -3768.

13. Vidali, M. Bioremediation: An overview. Pure and Applied Chemistry, 2001: 73(7), 1163-1173

14. Jorgensen, K.S., Puutstinen, J and Suortt, A. –M Bioremediation of petroleum hydrocarbon-contaminated soil by composting in biopiles. Environmental Pollution, 2000: 107, 245-254.

15. Department of Petroleum Resources. Environmental Guidelines and Standard for the Petroleum Industry in Nigeria, 2002

16. Joel, O.F and Amajuoyi, C.A Determination of selected physicochemical parameters and heavy metals in a drilling cutting dump site at Ezeogwu–Owaza, Nigeria. J. Appl. Sci. Environ. Manage, 2009: 13(2), 27- 31.

17. Okparanma, R.N., Ayotamuno, J. M Polycyclic aromatic hydrocarbons in Nigerian oil-based drill-cuttings; evidence of petrogenic and pyrogenic effects. World Applied Sciences Journal 2010; 11 (4): 394-400, ISSN 1818-4952.

18. Nweke, C.O and Okpokwasili, G. C Drilling fluid base oil biodegradation potential of a soil Staphylococcus species. African Journal of Biotechnology 2003; 2 (9), pp. 293-295. http://www.academicjournals.org/AJB

19. Ayotamuno, J.M., Okparanma, R, N and Araka, P.P Bioaugmentation and composting of oil-field drill-cuttings containing polycyclic aromatic hydrocarbons (PAHs). Journal of Food, Agriculture & Environment 2009; l.7 (2): 6 5 8 - 664. www.world-food.net

20. Okparanma, R.N Ayotamuno, J.M and Araka, P.P Bioremediation of hydrocarbon contaminated-oil field drill-cuttings with bacterial isolates. African Journal of Environmental Science and Technology 2009 3 (5), pp. 131-140. Available online at http://www.academicjournals.org/AJEST

21. Ifeadi, C.N The treatment of drill cuttings using dispersion by chemical reaction (DCR). A paper prepared for presentation at the DPR Health, Safety & Environment (HSE) International Conference on Oil and Gas Industry in Port Harcourt, Nigeria. 2004.

22. Adekunle, I.M., Ajijo, M.R., Omoniyi, I.T and Adeofun, C.O Response of four phytoplankton species in some sections of Nigeria coastal waters to crude oil in controlled ecosystem. Int. J. Environ., Res., Iran, 2009; 4 (1): 65 -74 http://ijer.ut.ac.ir

23. Adekunle, I.M and Onianwa, P.C Functional group characteristics of humic acid and fulvic acid extracted from some agricultural wastes. Nigerian Journal of Science, Nigeria, 2001: 35 (1), 15 – 19.

24. Adekunle, I.M Evaluating environmental impact from utilization of bulk composted wastes of Nigerian origin using laboratory extraction test. Environmental Engineering and Management Journal 2010; 9 (5): 721 -729.: http://omicron.ch.tuiasi.ro/EEMJ/

25. Adekunle I.M., Adekunle, A.A., Akintokun, A.K., Akintokun, P and Arowolo,T.A Recycling of organic wastes through composting for land applications: a Nigerian experience. Waste Management & Research 2011; 29 (6): 582 – 593. DOI: 10.1177/ http://wmr.sagepub.com/content/29/6/582.

26. Adekunle, I.M Bioremediation of soils contaminated with Nigerian petroleum products using composted municipal wastes. Bioremediation Journal, 2011; 15 (4): 230-241, DOI: 10.1080/10889868.2011.624137. http://dx.doi.org/10.1080/10889868.2011.624137

27. Adekunle I.M., Oguns, O., Shekwolo, P.D., Igbuku, O.O and Ogunkoya, O.O Assessment of population perception impact on value-added solid waste disposal in developing countries, a case study of Port Harcourt City, Nigeria. In: Xiao-Ying, Y (Ed) Municipal and Industrial Waste Disposal. Intech; 2012, p177-206.

28. Adekunle A. A., Adekunle, I.M., Igba, T. O Assessing the effect of bioremediation agent from local resource materials in Nigeria on soil pH. Journal of Emerging Trends in Engineering and Applied Sciences 2012; 3 (3) 526-532. http://jeteas. scholarlinkresearch.org/articles/Assessing%20the%20Effect%20 of%20Bioremediation%20Agent.pdf

29. Adekunle A.A., I.M. Adekunle and Igba, T.O Impact of bioremediation formulation from Nigeria local resource materials on moisture contents for soils contaminated with petroleum products. International Journal of Engineering Research and Development 2012; 2(4) 40-45 http://www.ijerd.com/paper/ vol2-issue4/F02044045.pdf

30. Adekunle A.A, Adekunle, I.M. and Igba, T.O Assessing and forecasting the impact of bioremediation product derived from Nigeria local raw materials on electrical conductivity of soils contaminated with petroleum products. Journal of Applied technology in Environmental Sanitation 2012; 2 (1) 57 -66. http:// www.trisanita.org/jates/atespaper2012/ates09v2n1y2012.pdf

31. Adekunle A.A., I. M. Adekunle and Igba T. O Soil temperature dynamics during bioremediation of petroleum products using remediation agent for Nigerian local resource materials. International Journal of Engineering Science and Technology 2012; 1 (4): 1-8. http://www.ijert.org/browse/june-2012-edition

32. Association of Official Analytical Chemists (AOAC), Official Method and Analysis of The Association oh The Official Analytical Chemists 11th Edition Washington D C, 1970.

33. Finar, I.L Organic Chemistry, volume I The Fundamental principles. 6th Ed, Longman, 1973.

34. Liu, W., Luo, Y and Teng, Y Bioremediation of oily sludge-contaminated soil by stimulating indigenous microbes. Environ Geochem health 2010: 32, 23 -29.

35. Ayotamuno, J.M., Okparanma, R.N., Davis, DD and allagoa, M. PAH removal from Nigerian oil-based drill-cuttings with spent oyster mushroom (Pleurotus ostretus) substrate. Journal of Food, Agriculture and Environment 2010: 8 (3 &4), 914 -919.

36. Rojas-Avelizapa, N.G., Roldan-Carrillo, T., Zegarra-Martinez, H., Munoz-Colunga, A.M and Fernadez-Linares A field trail for an ex-situ bioremediation of a drilling mud-polluted site. Chemospher, 2007: 66, 1595 – 1600.

37. Martin, J.A., Moreno, J.L., Hernandez, T and Garcia, C Bioremediation by composting of heavy oil refinery sludge in semiarid conditions. Biodegradation, 17:, 251 – 261.

38. Al-Nasrawi, H Biodegradation of Crude Oil by Fungi Isolated from Gulf of Mexico. J Bioremed Biodegrad 2012, 3:4

39. Mandal, A.K., Sarma, P. M., Singh, B., Jeyaseelan, C.P., Channasshettar, V.A., Lal, B and Datta, J bioremediation : an environment friendly sustainable biotechnological solution for remediation of petroleum hydrocarbon contaminated waste. ARPN Journal of Science and Technology, 2012: 2, 1-12

40. Stevenson, F.J Humus Chemistry, 2004. Wiley & Sons

41. Obayori, O.S., Ilori, M.O., Adebusoye, S.A., Amund, O.O and Oyetibo, G.O Microbial population changes in tropical agricultural soil experimentally contaminated with crude petroleum. African Journal of Biotechnology, 2008: 7 (24), 4512-4520.

42. Corwin, D.L and Lesch, S.M. Apparent soil electrical conductivity measurements in agriculture. Computers and Electronics in Agriculture, 2005: 46, 11–4

Chapter 4

Speed of Sound in Atmosphere of the Earth

Vladimir G. Kirtskhalia

Ilia Vekua Sukhumi Institute of Physics and Technology, Tbilisi, Georgia

ABSTRACT

It is demonstrated that contemporary conception on adiabaticity of sound in the Earth atmosphere is fair in sufficient approximation only for altitudes $z \leq 10^3$ m. At higher altitudes adiabaticity of sound is violated and essential dependence of its speed on altitude is revealed which is related to heterogeneity of the atmosphere in gravitation field of the Earth. It became possible to reveal the factor of gravity field due to the fact that in the equation of the state of atmosphere considered to be ideal gas, the entropy s is taken into consideration and is written down as ρ = (p, s) instead of generally accepted ρ = ρ(p) which is fair only for isentropic media and is not applicable to the Earth. Such approach enabled to determine that apart from adiabatic mechanism of generation of sound wave there exists isobaric one and exactly this mechanism leads to dependence of sound speed on altitude which is the same as dependence on density.

INTRODUCTION

Sound speed is a characteristic quantity of the medium included in the system of hydrodynamic (gas-dynamic) equations and plays a significant role in study of wave processes within it. Therefore, correct determination of its value is crucial in adequate description of generation and distribution of waves in media. According to the modern theory of sound wave density perturbation is considered as mass variation in constant volume without heat transfer i.e. adiabatically and therefore the speed of sound propagation is called adiabatic speed of sound. Pursuant to this representation speed of sound in atmosphere

is calculated according to the formula $C = \left(\gamma RT / M\right)^{1/2}$, i.e. depends only on temperature and does not depend on altitude. This result is dubious expressing that under the identical temperature conditions at the sea level and at an altitude of 100 km, where air density is 10^7 times less the speed of sound should have the same meaning. Figures 1 and 2 demonstrate graphs of dependence of atmosphere temperature of the Earth and sound speed at altitude in the range from of 0 to 85 km [1] which almost coincide. The same paper provides for the table demonstrating numerical meanings of sound speed at the altitudes of same interval and it is indicated that these meanings are calculated by means of the mentioned formula (Figure 13 shows the variation with altitude of the computed speed of sound). Internet provides for the same meaning of sound speed [2].

Such result is conditioned by incorrect definition of sound speed. As a matter of fact medium density may be Therefore sound may have not only adiabatic but isobaric speed as well. Combination of these two speeds must result in dependence of sound speed on altitude which constitutes qualitatively new result.

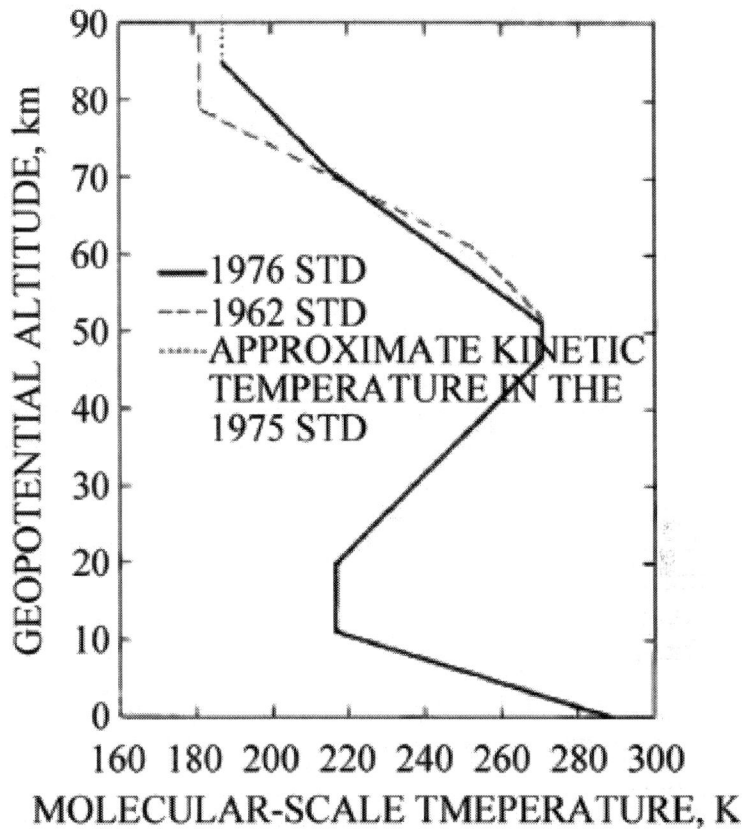

Figure 1: Molecular-scale temperature as a function of geopotencial altitude.

Urgency of this consideration is proved by the fact of lack of information on experimental measurements of dependence of sound speed on altitude carried out with sufficient accuracy. The paper [3] provides for the results of measurement of sound speed at the see level for conditions t = 0°C and p =1 atm and it is shown that it precisely coincides with theoretical results. perturbed not only as a result of change mass in constant volume but by change of constant mass volume during temperature vibration under the conditions of constant pressure i.e. isobaricly.

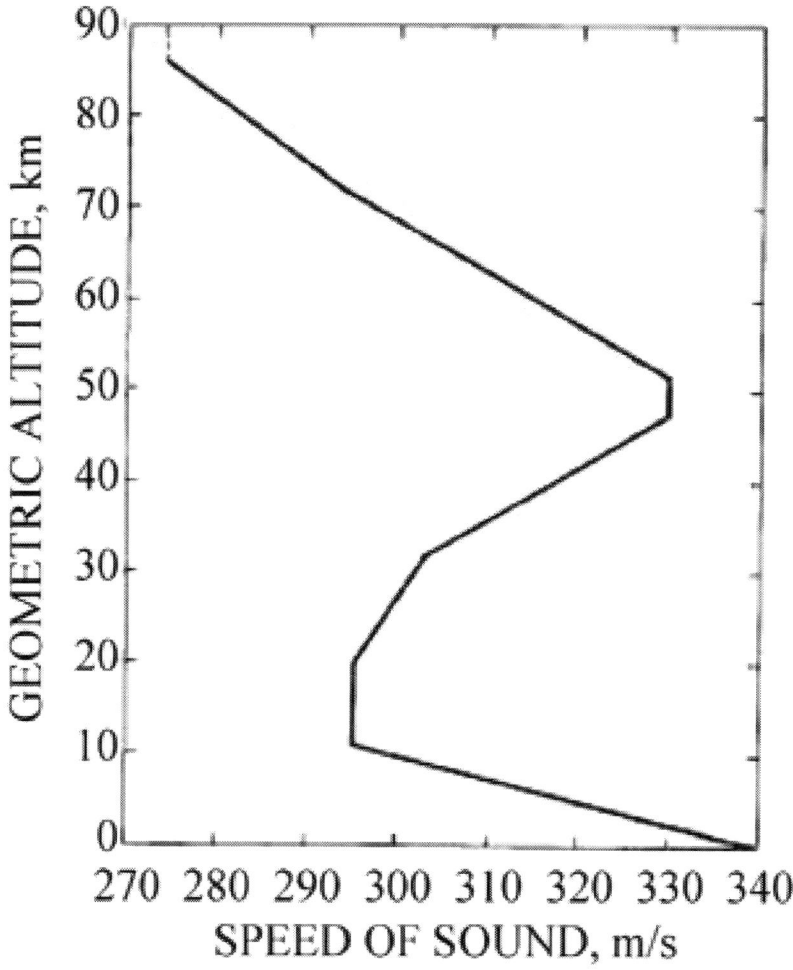

Figure 2: Speed of sound as function geometric altitude.

In work [4] the method and results of measurement of speed of sound in atmosphere of the Earth by means of a probe are given. The schedule of dependence of sound speed on time of vertical movement of probe is demonstrated which does not enable us to precisely determine dependence of sound speed on altitude. Notwithstanding the aforesaid, approximate analysis shows that the numerical results of experiment are below the existing theoretical data which is beneficial for us but exceed the results calculated by means of our theory. Therefore, it is required to check our theory through precise experiment.

The first section of this article presents critical analysis of two typical works of famous authors. It is shown that their theory of sound waves is fair only for homogeneous medium and is not applicable to the Earth atmosphere. In the second section it is demonstrated that taking into consideration of nonisentropicity of ideal gas existing in external force field leads to the result that the ideal value of square of the sound speed is reduced from squares of adiabatic and isobaric speed sounds

$\left(C^2 = C_s^2 / \left(C_s^2 + C_p^2\right)\right)$. The latter enables to provide correct definition of compressible and incompressible medium. In the third section it is shown that application of a new theory to the Earth atmosphere leads to dependence of sound speed in it on temperature as well as on altitude. A significant result is obtained evidencing that the upper bound of troposphere above which anomalous processes develop in atmosphere, coincides with the altitude where the values of adiabatic and isobaric speeds are equalized.

MODERN REPRESENTATIONS ABOUT SPEED OF A SOUND IN ATMOSPHERE OF THE EARTH AND INADEQUACY OF ITS APPLICATION

Motion of an ideal fluid (gas) in gravitational field of the Earth is described by the Euler's equation:

$$\rho\left[\frac{\partial \boldsymbol{v}}{\partial t} + \left(\boldsymbol{v} \cdot \nabla\right)\boldsymbol{v}\right] = -\nabla p + \rho \boldsymbol{g}$$

(1.1)

In wave motions, speed v is considered for small value. Pressure p and density r are represented in the form of the sum of their stationary (P_0(r), ρ_0(r)) and perturbation (p¢(r, t), r¢(r,t)) , and therefore after linearization of the Equation (1.1) we have:

$$\rho_0 \frac{\partial \boldsymbol{v}}{\partial t} = -\nabla\left(P_0 + p'\right) + \left(\rho_0 + \rho'\right)\boldsymbol{g}$$

(1.2)

Let's consider two examples of application of Equation (1.2) to sound waves which give clear representation of inadequacy of the current theory of sound waves in the Earth atmosphere.

In monographs [5,6] the Equation (1.2) is provided as follows:

$$\rho_0 \frac{\partial v}{\partial t} = -\nabla p'$$

(1.3)

here it is implied that g=0 and $\nabla p_0 = 0$ i.e. gravitational field does not affect the medium. This circumstance is highlighted in monograph [6], where it is indicated that under such approach the medium is isentropic, i.e. entropy s=const. The Equation (1.3) is solved together with the linearized equation of the continuity.

$$\frac{\partial \rho'}{\partial t} + \rho_0 (\nabla \cdot v) = 0$$

(1.4)

When relation between perturbations of pressure and density are given by the formula

$$p' = \left(\frac{\partial p}{\partial \rho} \right)_s \rho'$$

(1.5)

The index s means, that the derivative undertakes at constant entropy. Whereupon the potential of speed is defined as $v = \nabla \varphi$ with the Equations (1.3)-(1.5) that leads to the wave equation:

$$\frac{\partial^2 \varphi}{\partial t^2} - C_s^2 \Delta \varphi = 0$$

(1.6)

where quantity

$$C_s = \sqrt{\left(\frac{\partial p}{\partial \rho} \right)_s}$$

(1.7)

is defined as speed of adiabatic sound. Considering the air as ideal gas (p=nkT), and dependence between pressure and density in adiabatic process is defined by the following relation

$$\frac{p}{p_0} = \left(\frac{\rho}{\rho_0}\right)^{\gamma}$$

(1.8)

where p_0 and r_0 are initial values of pressure and density, and an adiabatic index γ– c_p/c_v=1.4 is ratio of thermal capacities (specific heath) for air in presence of constant pressure and volume respectively. From (1.7) for speed of sound following expression is derived:

$$C_s^2 = \gamma\frac{p}{\rho} = \gamma\frac{kT}{m_0} = \gamma\frac{RT}{M}$$

(1.9)

Here R = 8.314 J/(mol·K) is a gas constant, m_0 = 4.81·10^{-26} kg is mass of one air molecule and M = 29·10^{-3} kg/mol—mass of one mol air.

In monographies [7-9], the condition of static balance is applied to the Equation (1.2)

$$\nabla P_0 = \rho_0 \boldsymbol{g} ,$$

(1.10)

then it becomes:

$$\rho_0\frac{\partial v}{\partial t} = -\nabla p' + \rho' \boldsymbol{g}$$

(1.11)

Then it is assumed, that gravitational acceleration g does not influence high-frequency sound fluctuations and the Equation (1.11) passes in the Equation (1.3). The authors supplement the Equations (1.3) and (1.4) with the equation of adiabatic process for ideal gas

$$\frac{\partial p}{\partial t} - \gamma\frac{RT}{M}\frac{\partial \rho}{\partial t} = 0$$

(1.12)

and representing all perturbed quantities in the form of a plane wave [i(kx-wt)] reduce these three equations to the wave equation relative to x component of pretreated speed

$$\frac{\partial^2 u}{\partial t^2} - \gamma \frac{RT}{M} \frac{\partial^2 u}{\partial x^2} = 0$$

(1.13)

from which, they obtain expression (1.9) for the speed of a sound, which presently is applied for both bottom ($z \leq 11$ km) and top (11 km $\leq z \leq 86$ km) atmospheres.

Thus, in both cases authors try to define speed of a sound in atmosphere of the Earth from the wave equation for a plane wave with homogeneous phase speed, which depends only on temperature. It is well known, that to this equation satisfy intensity vectors of electric and magnetic fields of electromagnetic wave in vacuum. For which the vacuum is a homogeneous environment for its, consequently the velocity of light in it is homogeneous. Earth atmosphere is essentially heterogeneous medium for sound where all characteristic thermodynamic parameters depend on z coordinate. Since the sound of speed is also characteristic value of the medium, according to the fundamental principle of physics it must depend on z coordinate. Proceeding from the aforementioned, formulation and solution of the problem in the monographs [5,6] look correct, however they have no relation to a sound wave in atmosphere of the Earth. As for the resolution obtained in monographs [7-9], it can be easily shown that it is incorrect. Actually, in Euler's linearized equation the authors apply condition of statistic balance (1.10) where

$$P_0 = P_0^0 - \rho_0^0 g z .$$

(1.14)

Here P_0^0 and ρ_0^0 is values of pressure and density on a sea level. Expression (1.14) turns out from the general formula of distribution of pressure in atmosphere of the Earth

$$P_0 = n_0 kT = \frac{\rho_0}{m_0} kT = \frac{\rho_0^0}{m_0} kT \exp\left(-\frac{m_0 g z}{kT}\right),$$

(1.15)

where Laplace's barometric formula for ideal gas is used

$$\rho_0 = \rho_0^0 \exp\left(-\frac{m_0 g z}{kT}\right).$$

(1.16)

Assuming $m_0 gz/kT < 1$ and representing (1.16) in series, we obtain

(1.14) from (1.15) in linear approach. Thus, expression (1.14) is valid for the altitudes that fulfill inequality

$$\frac{m_0 gz}{kT} \leq 10^{-1}$$

(1.17)

Assuming that T = 288.15 °K on a sea level, we find, that z ≤ 850 m. Thus, the Equation (1.14) is applicable in atmosphere only up to altitude of 850 m, or in water [9] where density practically does not depend on depth.

Besides, analysis of available literature obviously reveals one more significant contradiction related to determination of sound speed in the Earth's atmosphere. In papers [10,11] the authors consider phenomena of sound dispersion and absorption. They deem that adiabatic speed of sound expressed by formula (1.9) makes no influence on the mentioned process and therefore, this speed of sound is called speed of sound of zero frequency. It is clear that this is a conventional name however it contradicts to the opinion of the authors of papers [6-9] stating that this speed complies with high frequency speed. The typical examples resulted above and ours comments show, that the existing theory of sound waves in atmosphere of the Earth demands revision.

THE REDUCED SPEED OF SOUND

The sound wave transfers density perturbation at mechanical displacement of particles of medium. Pursuant to understanding established at present, atmosphere air represents ideal gas density of which depends only on pressure, i.e.$\rho = \rho(p)$ In paper [6] it is noted that such approach is justified only for homogeneous medium in every point of which entropy has the same meaning (isentropic medium). It is apparent that the Earth's atmosphere does not meet this condition and therefore.$\rho = \rho(p,s)$. In terms of the aforementioned density perturbation should be put down as:

$$\rho' = \left(\frac{\partial \rho_0}{\partial P_0} \right)_s p' + \left(\frac{\partial \rho_0}{\partial S_0} \right)_p s'$$

(2.1)

The first member in (2.1.) corresponds to density perturbation caused by mass alteration in fixed volume as a result of pressure perturbation in conditions of constant entropy, while the second member corresponds by volume alteration of fixed mass as a result of entropy perturbation in conditions of constant pressure. On the other hand, entropy perturbation is also callused by pressure perturbation, i.e.

$$s' = \left(\frac{\partial S_0}{\partial P_o} \right)_T p'$$
(2.2)

(2.1) and (2.2) give

$$\rho' = \frac{1}{C^2} p'$$
(2.3)

Where

$$C^2 = \frac{C_S^2 C_P^2}{C_S^2 + C_P^2}$$
(2.4)

$$C_S^2 = \left(\frac{\partial P_0}{\partial \rho_0} \right)_S$$
(2.5)

$$C_P^2 = \left[\left(\frac{\partial \rho_0}{\partial S_0} \right)_P \left(\frac{\partial S_0}{\partial P_0} \right)_T \right]^{-1}$$
(2.6)

If expression (2.5) is called adiabatic speed of sound, it is logical to call expression (2.6) isobaric speed of sound since under thermodynamic relations

$$\left(\frac{\partial \rho_0}{\partial S_0} \right)_P = \frac{T}{c_P} \left(\frac{\partial \rho_0}{\partial T} \right)_\rho \quad \text{and} \quad \left(\frac{\partial S_0}{\partial P_0} \right)_T = \frac{1}{\rho_0^2} \left(\frac{\partial \rho_0}{\partial T} \right)_P$$

application for (2.6) results in

$$C_P^2 = \frac{c\,\rho_0^2}{T\left(\dfrac{\partial \rho_0}{\partial T}\right)_P^2}$$

(2.7)

(15) demonstrates that in homogeneous Medium Square of sound speed coincides with coefficient of connection of pressure and density perturbations. We assume that the same is fair for heterogeneous medium and therefore it should be defined by means of formula (2.4). We see that square of ideal value of sound speed equals to the reduced value of square of adiabatic and isobaric speeds. Such definition of sound speed—which is fair for any medium in the field of gravity, profoundly changes the existing notion on the sound itself and the medium. Pursuant to the latter, medium is considered incompressible if speed of sound in it is $C=C_s= \infty$ At such definition of the incompressible medium, which model is often used in applied problems, equation of medium state (2.3) is fell out of the system of hydrodynamic (gas dynamic) equations and thus, it is not clear which medium is referred. Deriving from (2.4) medium is considered incompressible, if $C_s > C_p$ and then, $C \approx C_p$ and for the compressed medium $C_p > C_s$ and $C \approx C_s$. Now, the concept of compressibility or incompressibility gets clear physical sense that has significant applied value.

SPEED OF SOUND IN ATMOSPHERE OF THE EARTH

Atmosphere of the Earth represents multilayered structure and in each layer, dependences of physical parameters on geometrical altitude z are different. The chart provided for in Pic. 1, which can be found in internet [2] as well as in scientific literature [1;7] shows that in interval of altitudes from z = 0 to z = 11 km temperature changes under linear law T=-6.52.10^{-3}z +288.15 and in the interval from z = 51 km to z =

85 km approximately $T = -2.60.10^{-3}(z - 51.10^3) + 270.50$ under the law. In intermediate interval (11 - 51 km) the temperature is either constant or increases, which makes us think that anomalous processes take place there. Besides, along with increase of altitude, deviation of experimental value to a greater index from the values calculated by means of formula (1.16) is also flaring. E.g. at the altitude $z = 7.5$ km where theoretical value of air density decreases e-times, relative error constitutes 23%, while when $z = 11$ km it constitutes 40%. At 35 km they equalize and thereafter theoretical value exceeds the experimental one and at the altitude of 85 km relative error constitutes 95%. In light of the aforementioned we assume that the model of ideal gas and consequently our theory can be applied up to 11 km altitude where temperature fall strictly obeys linear law. The aforesaid does not exclude possibility of its application in conditions of upper atmosphere.

Let's demonstrate that theoretical calculations shown above result in obvious dependence of sound speed on altitude in the Earth's atmosphere. Indeed, for value C_s we have expression (1.9) while by application of (1.16) from (2.7) for C_p we get

$$C_p = \sqrt{\frac{c_p k^2 T^3}{m_0^2 g^2 z^2}}$$

(3.1)

Substituting (1.9) and (3.1) in (2.4) expressions for speed of a sound in the specified layer ($0 \leq z \leq 11$ km) of the Earth atmosphere we receive

$$C = \sqrt{\frac{\gamma k T}{m_0 \left(1 + \dfrac{\gamma m_0 g^2 z^2}{c_p k T^2}\right)}}$$

(3.2)

(3.2) shows that ideal value of sound speed in atmosphere obviously depends on altitude. Besides, we see that sound is adiabatic not only when $g = 0$ but when $g \neq 0$ but $z = 0$. At removal from a sea level, dependence of speed of sound on altitude becomes evident which is caused by heterogeneity of the atmosphere. In Table 1, C_s, C_p and C values are obtained by formulas (1.9), (3.1) and (3.2) in the range of altitudes from to $z=0$ up to $z=11$ km, besides, corresponding values of speeds of sound C_{int} taken from the online calculator [2]. Values C_s and

C_{int} coincide, and relative errors between values C and C_{int} at altitudes of z=1 km and z = 11 km are equal to 0.3% and 33% respectively. It confirms our assumption that it is possible to consider sound to be adiabatic only up to altitude of $z \leq 850$ m. For visualization, dependences of C_s and C on z in intervals (0 - 1 km) and (1 - 11 km) are presented on the Figures 3 and 4 respectively. Let's define the altitude on which $C_s = C_p$ From (3.2) it can be seen, that it is defined from a relation Substituting values of constants ($c_p = 10^3$ J/kgK) in (3.3) and assuming that average value of temperature in the mentioned interval is T = 252.4 °K, we receive z = 11.6 km. As we can see altitude on which adiabatic and isobaric speeds of sounds are equal, practically precisely coincides with altitude above which anomalistic processes occur in atmosphere. We think that it is not a casual coincidence and it has deep physical sense, definition of which is a subject of the further researches.

$$z = \sqrt{\frac{c_p k \, \overline{T}}{\gamma m_0 \, g}}$$

(3.3)

Table 1: Velocity values of C_s, C_p and C at altitudes from 0 too 11,000 m; comparison with internet data (online calculator)

Z(m)	T(k)	C_{int}(m/s)	C_s(m/s)	C_p (m/s)	C (m/s)
0	288.15	340.29	340.30	infinity	340.30
50.00	287.82	340.10	340.11	90180.02	340.11
100.00	287.50	339.91	339.92	45013.66	339.91
200.00	286.85	339.53	339.53	22430.54	339.50
300.00	286.20	399.14	339.15	14902.90	339.06

400.00	285.55	338.76	338.76	11139.12	338.61
500.00	284.90	338.38	3.38.38	8880.88	338.13
600.00	284.25	337.98	337.99	7375.42	337.64
700.00	25 3.60	*337.60*	337.61	6300.12	337.12
800.00	282.95	337.21	337.22	5493.66	336.59
900.00	282.30	336.82	336.83	4866.44	336.03
1000.00	281.65	336.43	336.44	4364.68	335.45
2000.00	275.15	332.53	33'⁷.54	2107.23	328.47
3000.00	268.65	328.58	328.59	1355.33	319.34
4000.00	262.15	324.58	324.59	979.83	308.12
5000.00	255.65	320.53	320.54	754.89	295.04
6000.00	249.15	316.43	316.44	605.24	280.42
7000.00	242.65	312.27	312.28	498.61	264.66
8000.00	236.15	308.06	308.07	418.87	248.18
9000.00	229.65	303.79	303.80	357.06	231.3S
10000.00	2/3.15	299.46	299.47	307.81	214.65
11000.00	216.65	295.07	295.08	267.69	198.26

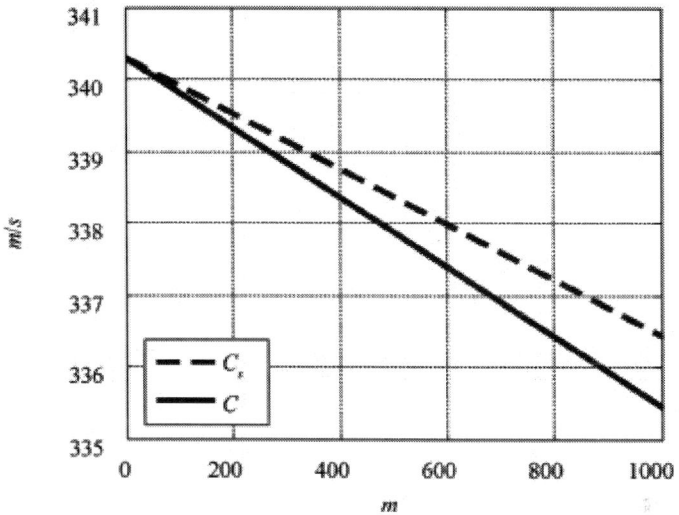

Figure 3: Dependence of C_s and C on altitude in the range z = 0 m to z = 1000 m.

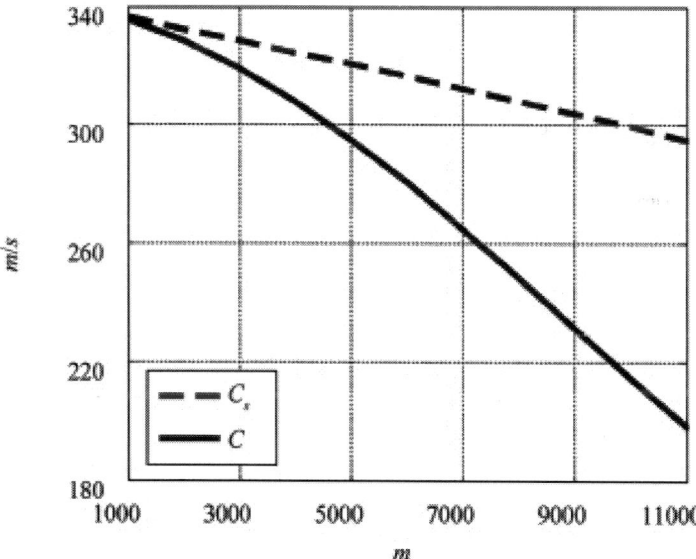

Figure 4: Dependence of C_s and C on altitude in the range z = 1000 m to z = 11,000 m.

CONCLUSIONS

It is shown above that the existing theory on adiabatic sound in the Earth atmosphere is characterized by deficiencies related to incorrect definition of the sound speed which is understood as speed of distribution of adiabatic density perturbation of the medium. As a result, the sound speed in the Earth atmosphere depends only on temperature and does not depend on altitude while the density itself exponentially depends on it. As a matter of fact the density may be perturbed not only adiabaticaly, but isobaricaly as a result of entropy perturbation at perturbation of pressure leading to perturbation of temperature and consequently to isobaric alteration of the volume of the given air mass. All this finally results in dependence of sound speed on altitude pursuant to the formula (3.2). We assume that correction of the adiabatic wave theory suggested by us has an unconditional right to exist unless it is disproved by an experiment. However, if it is proved it will lay the foundation for qualitatively new research in gas and hydrodynamics.

REFERENCES

1.   U.S. Standard Atmosphere, National Aeronautics and Space Administration, 1976.

2.   http://www.digitaldutch.com/atmoscalc

3.   G. S. K. Wang, "Speed of Sound in Standard Air," Journal of the Acoustical Society of America, Vol. 79, No. 5, 1986, pp. 1359-1366. doi:10.1121/1.393664

4.   G. Santostasi, et al., "A Student Designed Experiment Measuring the Speed of Sound as a Function of Altitude," McNeese State University, Lake Charles, 2008.

5.   L. D. Landau and E. N. Lifshitz "Nauka," Theoretikal Physics, Hydrodynamics, Vol. 6, Moscow, 1988.

6.   A. D. Pirce, "Acoustics: An Introduction to Its Physical Principles and Applications," Acoustical Society of America, New York, 1989.

7.   E. E. Gossard and W. H. Hooke, "Waves in the Atmosphere," Elsevier, New York, 1975.

8.   L. M. B. C. Campos, "On Three-Dimensional Acoustic Gravity Waves in Model Non-Isothermal Atmospheres," Wave Motion, Vol. 5, No. 1, 1983, pp. 1-14. doi:10.1016/0165-2125(83)90002-1

9.   M. J. Lighthill, "Waves in Fluids," Cambridge University Press, Cambridge, 2002.

Chapter 5

Assessing Landform Alterations Induced by Mountaintop Mining

Aaron E. Maxwell[1] and Michael P. Strager[2]

[1]Natural Resource Analysis Center, West Virginia University, Morgantown, USA

[2]Division of Resource Management, West Virginia University, Morgantown, USA

ABSTRACT

A comprehensive impact analysis of mountaintop removal and valley fill (MTR/VF) mining requires an understanding of landform alterations since ecological impacts are so intricately linked. In this study we investigated mining in the Coal River Watershed, West Virginia, USA, using landform terrain analysis. Previous studies have relied on elevation differencing of preand postmining surfaces to assess absolute elevation and volumetric change. Our landscape analysis, utilizing light detection and ranging (LiDAR)- derived elevation data, indicated specific landform types and distributions that were significantly altered after MTR/VF mining and reclamation. The use of categorical landform data provides insights to assessing and understanding the extent of topographically altered mountaintops. Our study provides an opportunity to further examine the impact of MTR/VF on forest communities, terrestrial habitat, ecosystem health, and biodiversity.

INTRODUCTION

In the Southern Coalfields of West Virginia, MTR/VF is currently the leading cause of land cover change [1-4]. Multiple watersheds in West Virginia have more than 10% of their surface area disturbed by surface mining [5], which results in the loss of forest and a conversion to barren land cover [6]. It has been estimated that all surface mining in Appalachia has resulted in a net loss of 420,000 ha of forest [3], and the interior character of the forest is threatened by the introduction of non-forest edges [7]. During the MTR/VF process, forests are cleared, top soil is removed, overburden material is blasted away to uncover coal seams, and overburden material is placed in adjacent valleys, filling stream segments and creating valley fills [5,8]. In Kentucky, it has been estimated that greater than 660 km of headwater streams were buried between 1985 and 1999 [9]. Later reclamation produces grasslands or shrub/scrub land cover; however, productivity is often limited due to poor soil conditions [6]. Attempts are often made to preserve soil; however, it becomes homogenized and soil horizons are not maintained [10].

In addition to the land use and land cover (LULC) change, the landscape and terrain are recontoured with modified watershed ridges, altered vegetation conditions, and modified soil character [8]. References [1,11] estimated that surface mining was responsible for displacing more material in the Southern Coalfields of West Virginia than river systems and natural geomorphic processes. Furthermore, MTR/VF methods have resulted in more moved material and faster landscape alterations as compared to traditional surface mining methods, such as auger, contour, and highwall mining [12].

Elevation differencing of Digital Elevation Models (DEMs) representing different temporal conditions has been explored to describe topographic change. For example, [13] utilized NASA Shuttle Radar Topographic Mission (SRTM), US Geologic Survey National Elevation Dataset (NED), and Terrain Resource Information Management Program (TRIM) data to describe and map alpine glacier changes in southeast Alaska and northwest British Columbia. Reference [14] utilized LiDAR-derived DEMs to calculate dune volume changes over a 1-year period at sites along the Cape Hatteras National Seashore.

Landscape and geomorphic change resulting from MTR/VF disturbance were specifically analyzed in Perry County, Kentucky, USA [15] utilizing NED (pre-mining) and SRTM (post-mining) data. The study highlighted the complexity of such an analysis when timestamps of the NED data are variable since they were created relative to best available data and 1:24000 scale topographic quadrangle maps. In a similar study, the West Virginia Department of Environmental Protection (WVDEP) utilized interferometric synthetic aperture radar (IFSAR) and elevation raster data produced from digital line graph (DLG) hypsography to map valley fill extents throughout nine counties in southern West Virginia. Because the radar data did not adequately penetrate the tree canopy, it was necessary to remove forested areas from the analysis. It was found that a complete inventory of the fills required additional visual classification [16].

To the best of our knowledge no previous work has quantified the post-mining landscape in terms of changes in terrain characteristics from pre-mining conditions using landscape-scale categorical terrain data. An understanding of the terrain alteration using landform data is appropriate to assess the impact of the topographically altered mountaintops.

This paper expands upon earlier differencing and topographic change work in the MTR/VF region by implementing a methodology relying on a categorical representation of the landscape as landforms. This data differentiates the landscape into the following classes: cliff, steep slope, slope crest, upper slope, flat summit, sideslope, cove, dry flat, moist flat, wet flat, and slope bottom. This method was adopted after [17] because such features adequately represent the Southern Coalfields at the landscape scale.

Attempts have been made to link landform data to habitats using predictive modeling. For example, [17] attempted to link the ecological community types developed by the Nature Conservancy to landforms, elevation, and lithology. These categories represent landforms of ecological significance at the landscape scale. Our goal was to investigate how the distribution of these categories was impacted by surface mining. Our results may help to reestablish a post-mining landscape that can benefit terrestrial habitat, ecosystem health, and biodiversity since these have been shown to be linked to landform clasess [17].

METHODOLOGY

Study Area

The Coal River Watershed is a 230,755 ha Hydrologic Unit Code (HUC) 8 headwater watershed completely within West Virginia, USA. The Coal River Watershed exists within the Appalachian Plateau physiogeographic province, a dissected, westward-tilted plateau dominated by Pennsylvanian bedrock. Pennsylvanian stratigraphy is characterized as cyclic sequences of sandstone, shale, clay, coal, and limestone [18]. The terrain is dissected by a dendritic stream network and shows fine texture with moderate to strong local relief. In comparison to the northern Appalachian Plateau, the Southern Coalfields is generally more rugged due to resistant strata [19]. The terrain analysis using landforms suggests that this terrain is naturally dominated by steep slopes.

According to LULC estimates derived from aerial photography, 8.8% of this watershed was disturbed by active surface mining or mine reclamation in 2009. It should be noted that this estimate does not take into account historical mining areas which have since been reforested. Figure 1 shows the watershed location within West Virginia. This watershed was selected as a case study because it is heavily impacted by mining, there has been continued mining and reclamation between 2003 and 2010, and because LiDAR data were available for the extent.

Digital Elevation Data Utilized

The LiDAR data, representing recent topographic conditions, were collected between April 9th, 2010 and April 18th, 2010 at a flight height of 1524 meters above ground level (AGL), a pulse frequency of 70 kHz, a scan frequency of 35 Hz, and a scan angle (full field of view) of 36˚. The swaths were flown with a 30% overlap, at an average speed of 250 km/hr, and with an average width of 979 m. An Optech ALTM 3100 C sensor was used to collect the data. The scan and flight specifications were selected to support a 0.7 m contour interval and a 1 m nominal ground post spacing. Ground points were filtered from the all return data to produce ground point files in LAS 1.2 format.

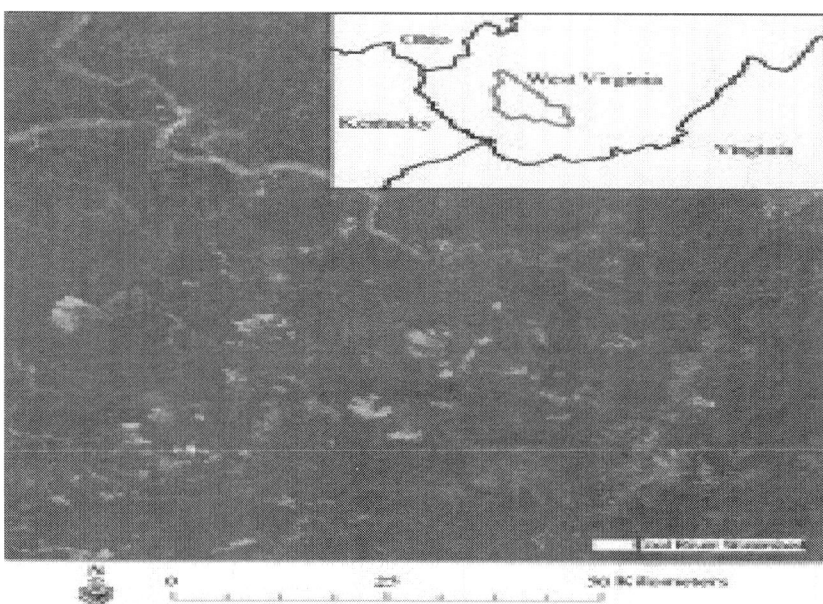

Figure 1: Study Area.

An interpolated raster surface was created from point data using ArcMap 10, and average point spacing of 0.01 m. Inverse distance weighting (IDW) was used to interpolate a raster surface since sample points were in an evenly distributed pattern. This process resulted in a 1 m resolution, floating point elevation dataset.

Assessing Landform Changes between 2003 and 2010

Landform changes were assessed in filled and cut or excavated areas resulting from mining and reclamation between 2003 and 2010. The 2010 LiDAR-derived DEM was compared to a 2003 photogrammetrically-derived DEM representing 2003 conditions. The 2003 DEM data were provided by the West Virginia Statewide Addressing and Mapping Board (SAMB) who contracted BAE Systems ADR to create a stereo photogrammetric-derived digital terrain model (DTM) from statewide, spring, leaf-off, 1:4800 scale aerial photography. The DEM was created in compliance with National Dataset standards (1/9th arc second) and produced from mass points and breaklines. This dataset supports a vertical accuracy of +/− 3.048 m, root mean square error (RMSE) and has a cell size of 3 m [20].

In order to match the resolution of the SAMB DEM, we resampled the LiDAR-derived raster to 3 m resolution using bilinear interpolation and snapped the grid to match the extent of the SAMB DEM. This process assured that the raster grids were completely aligned. This process resulted in two 3 m DEMs representing 2003 and 2010 conditions that could be compared. A vertical data transformation was unnecessary since both DEMs had a vertical reference of NAVD88 orthometric.

In order to detect systematic error between the 2003 and 2010 DEM data, a total of 281 points were collected in the field throughout the Coal River Watershed on flat, paved surfaces using Pacific Crest realtime kinematic (RTK) survey equipment. The elevation measurements from the 2003 and 2010 elevation raster data were obtained at these locations. The measurements from each DEM were then compared. The mean difference was −0.4 m with a maximum difference of 0.8 m and a minimum difference of −2.0 m. Based on this analysis we did not correct for systematic difference between DEMs.

Once DEMs were obtained, processed, and prepared, they were subtracted using Raster Calculator within the Spatial Analyst Extension of ArcMap [21]. This produced a grid of elevation differences throughout the watershed. Negative values indicated potential cuts or excavations while positive values indicated potential fills. However, difference could simply have resulted from errors in the DEMs or in methodology. As a result, once this elevation difference model was calculated, it was necessary to determine a tolerance or threshold that would constitute true change and not simply error or noise between the digital elevation datasets. The photogrammetric DEM had an error tolerance (RMSE) of +/− 3.048 m while the LiDAR DEM had and error tolerance of only 15 cm (0.15 m). An equation suggested by [22] was used to estimate this threshold. The RMSE of each dataset was squared, the results were summed, and the square root of the sum was then taken. The result was then multiplied by three to determine a cut off representing values that were greater than three standard deviations from the mean. This method assumes a Gaussian distribution. Elevation change measurements outside of this range were considered true elevation differences and not a result of error or noise. Reference [15] used the same methodology to derive an error tolerance for the analysis conducted in Perry County, Kentucky, USA.

According to this method, it is not certain whether differences less than +/− 9.2 m represented true topographic change. As a result, the elevation difference grid was reclassified so that values between −10 m and +10 m were considered no change, error between the DEMs, or noise. Pixels with values less than −10 m were considered potential cuts while pixels with values greater than 10 m were considered potential fills. This process produced a cut and fill mask.

A LULC dataset was created for the region from 2009 National Agriculture Imagery Program (NAIP) orthophotography. Optimally, imagery collected at the time of the LiDAR collection would have been utilized; however, such data were not available. The imagery was classified using an object-based feature extraction methodology, augmented with GIS decision rules and manual digitizing. Forest, grass, and barren land cover were extracted from the raw imagery.

Combining these data with mine permit boundaries from the West Virginia Department of Environmental Protection (WVDEP) made it possible to delineate barren and grassland land cover in mine permits. Grasslands in permits represent potential reclamation while barren areas represent areas of active mining or areas that have not yet been reclaimed. The extents of valley fill faces and slurry impoundments were digitized. According to an error assessment based on manual aerial photograph interpretation at 100 randomly selected points, the LULC dataset had an overall accuracy of 94% (K_{Hat} of 93%).

It was possible to further remove erroneous pixels as potential cut or fills by utilizing additional data, such as the high resolution land cover. For example, potential cut or fill pixels could be removed if they were outside of disturbance resulting from surface mining. The potential cut or fill pixels that existed within areas of mine disturbance (barren, grasslands, or valley fill faces) were considered. Slurry impoundments were excluded because of errors associated with water level. To complete this analysis, it was necessary to convert the raster data to polygons that represented contiguous areas of cut or fill. No smoothing or simplification was applied.

Small areas of topographic change could also be removed as error or noise. Thresholds were determined from the hand digitized valley fill faces. The average size of these faces was 73,257 m^2(7.3 ha) with a standard deviation of 67,013 m^2 (6.7 ha). Based from this distribution, any contiguous area of cut or fill that was smaller than 60,000 m^2 (6 ha) was considered noise and removed. Although this threshold was somewhat arbitrary, it was selected since this study focused on detecting large expanses of excavation and fills.

The result of this analysis was a vector layer of potential cut or fill extents in which topographic change had occurred between 2003 and 2010. The extents were used to define areas of potential terrain change for analysis and comparison of the distributions of landforms. Although there was error associated with this methodology, it was adequate for this analysis since we were not interested in conducting an assessment of absolute elevation change or volumetric change but in defining areas in which to compare categorical terrain data and their impact on forest communities.

Creation of Landforms

The DEM data were used to classify the terrain into the following landforms: cliff, steep slope, slope crest, upper slope, flat summit, sideslope, cove, dry flat, moist flat, wet flat, and slope bottom. The methodology described by [17] provides a means to utilize DEM data to classify the landscape into different units of ecological significance. DEM derivatives including slope in degrees, a hydrologically filled DEM, flow direction, and flow accumulation were created using ArcMap 10. The slope and flow accumulation grids were utilized to calculate a moisture index. The moisture index is a relative measure of the moisture of a specific cell, and it assumes that the moisture level is a function of how much water flows into the cell, predicted by flow accumulation, and how fast the water can flow out, described by slope [20-25].

Landscape position was calculated following [26] to divide the landscape into the following categories: ridge, wide ridge, slope/flat, slope/cove. The approach is based on a local neighborhood analysis of elevation values in which a cell is compared to the mean elevation of all values within 3 by 3 windows. Landscape position was combined with slope and moisture index data to derive the final landform classes. This procedure was conducted for both the 2003 and the 2010 DEMs. The comparisons of the derived landforms included:

- Areas filled between 2003 and 2010 (mining related);
- Areas cut or excavated between 2003 and 2010 (mining related);
- Areas of mine reclamation in 2009, areas of mining disturbance or reclamation in 2009, and areas in WVDEP mine permits but not disturbed (forested in 2009).

This enabled the comparison of post-mining landform categories with pre-mining conditions.

RESULTS

Figure 2 shows the distribution of landforms in a MTR/VF site representing 2010 conditions. Comparing the mine site with the surrounding landscape, the distribution of landforms is greatly altered by mining even after reclamation has occurred. Figure 3 shows a different MTR/VF site and offers a comparison of landforms within areas cut or filled between 2003 and 2010, and this visualization also shows a redistribution of landforms.

Landforms were compared for areas of cut or fill between 2003 and 2009 within mine disturbance. Such areas represent potential areas of true topographic change. The results are described in Tables 1 and 2 and Figures 4 and 5. Chi-square tests were performed on the categorical data for this observational study, and the results suggest a statistically different distribution of landforms after volumetric change ($\alpha = 0.001$).

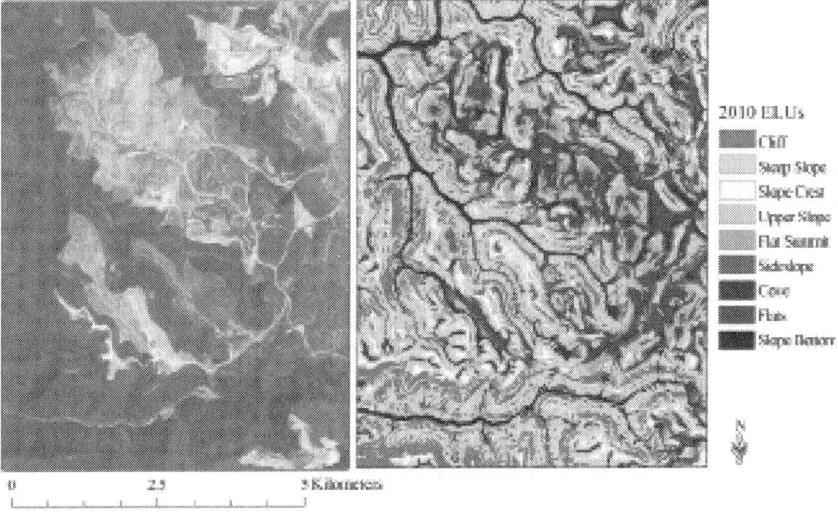

Figure 2: Example of landform distribution after landscape alteration and MTR/VF. The imagery is orthophotography representing growing season 2011 conditions.

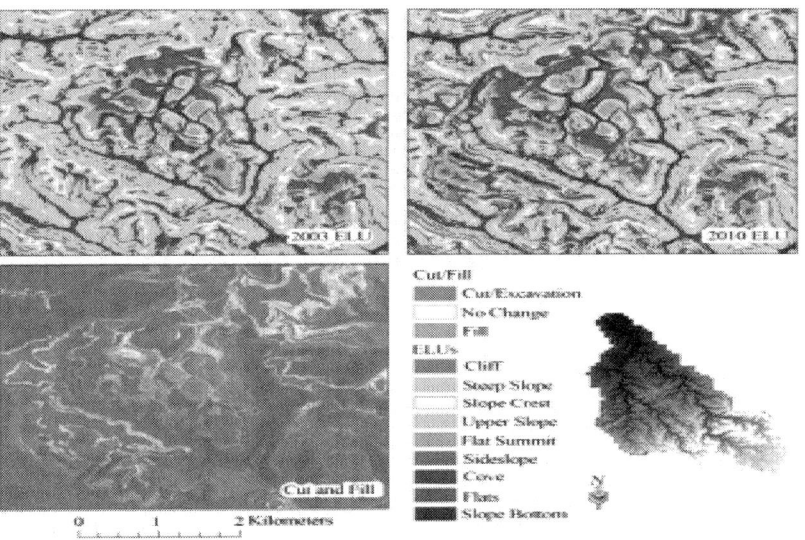

Figure 3: Comparison of landforms at mine site. The imagery is orthophotography representing growing season 2011 conditions.

It was possible that differences in the DEM production methodology could have resulted in differences in the resulting landform classification. This was observed by visual inspection of the results and was a source of error in the data represented in Figures 4 and 5. As a result, differences in classification and distribution of landforms could result from true terrain change or could be a product of the differences in DEM production methodology, photogrammetric or LiDAR-derived. In order to resolve such ambiguity, terrain classification differences were assessed for different surface mining LULC categories (2009) using the 2010 ELU data only. The results are described in Table 3 and Figure 6 in which landform classes were compared for areas in mine permits but still forested or not disturbed, areas of mine reclamation, and areas of mine reclamation or mining disturbance. The assumption was that areas permitted for surface mining should generally have similar topography. As a result, undisturbed areas within the mine permits representing pre-mining terrain were compared to mined or reclaimed areas representing post-mining terrain. This analysis showed a marked change in landforms from steep slopes to more flat topography, such as moist flats, upper slopes, and flat summits. We argue this, this is a more valid method to assess topographic change induced by surface mining and excavation because this technique was not impacted by differences in DEM methodology.

A chi-square test was performed to test for a statistical difference in the distributions of landforms associated with Table 3 and Figure 6. The test suggest a statistically significantly different categorical distribution of landforms pre and post-mining ($\alpha = 0.001$). The landform classes were then reclassified to only slopes and flats, and a second chi-square test suggests a statistically significant increase in flats and decrease in slopes postmining ($\alpha = 0.001$).

Two sample Student's t-tests were used to determine if pre and post-mining average slope conditions were statistically different. A comparison of slopes for all pixels in areas that were not mined in 2003 but were predicted as reclaimed relative to summer 2009 conditions showed a statistically significant different average slope at the 99.9% confidence level. However, as noted above, difference could have been a result of different methodol-ogy utilized to produce the DEM data. As a result, we also compared all pixels that were forested in permits, relative to growing season 2009 conditions, and areas that were either reclaimed or mined disturbed or reclaimed.

Table 1. Distribution of landforms for areas within mines and filled between 2003 and 2010

ELU	2003 Ha	Percent	2010 Ha	Percent
Cliff	257.34	9.20%	90.97	3.25%
Steep Slope	1128.87	40.34%	444.05	15.87%
Slope Crest	14.16	0.51%	84.20	3.01%
Upper Slope	101.72272	3.63%	413.88	14.79%
Flat Summit	66.14	2.36%	379.32	13.55%
Side Slope	291.75	10.43%	568.09	20.30%
Cove	602.42	21.53%	202.18	7.22%
Dry Flat	3.00	0.11%	76.29	2.73%
Moist Flat	106.85	3.82%	427.81	15.29%
Wet Flat	82.40	2.94%	34.46	1.23%
Slope Bottom	143.80	5.14%	77.15	2.76%
TOTAL	2798.45	100%	2798.40	100%

Table 2. Distribution of landforms for areas within mines and cut or excavated between 2003 and 2010

		2003	2010	
ELU	**Ha**	**Percent**	**Ha**	**Percent**
Cliff	261.11	10.59%	214.42	8.69%
Steep Slope	1025.26	41.57%	314.84	12.76%
Slope Crest	564.54	22.89%	134.42	5.45%
Upper Slope	280.71	11.38%	419.97	17.03%
Flat Summit	169.72	6.88%	446.11	18.09%
Side Slope	96.98	3.93%	364.3	14.76%
Cove	41.38	1.68%	165.93	6.73%
Dry Flat	1.04	0.04%	56.22	2.28%
Moist Flat	12.07	0.49%	212.46	8.61%
Wet Flat	11.01	0A5%	20.89	0.85%
Slope Bottom	2.76	0.11%	117.29	4.76%
TOTAL	2466.58	100%	2466.58	100%

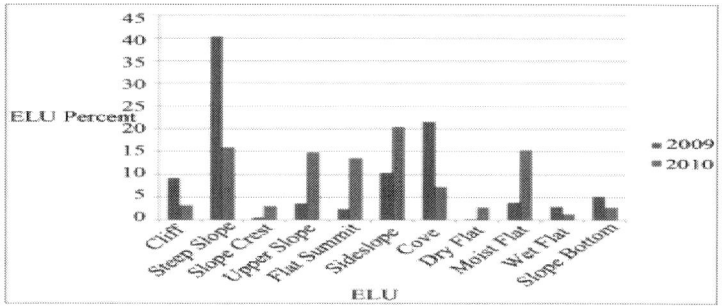

Figure 4: Comparison of landforms for areas within mines and filled between 2003 and 2010.

Table 3. Distribution of landforms within different land cover categories (2009) relative to 2010 DEM

ELU	Forest in Permit Ha	Reclaimed Ha	Mine Ha
Cliff	3306.74	596.99	1165.89
Steep Slope	11328.50	2777.78	4274.67
Slope Crest	1939.41	909.98	1443.73
Upper Slope	1638.62	2042.73	3413.90
Flat Summit	464.11	1277.28	2697.18
Side Slope	1587.01	2262.82	3723.33
Cove	2493.24	1388.31	2268.33
Dry Flat	75.05	152.48	406.80

Moist Flat	105.30	887.71	2015.08
Wet Flat	11.87	76.44	174.32
Slope Bottom	418.58	488.21	1040.06

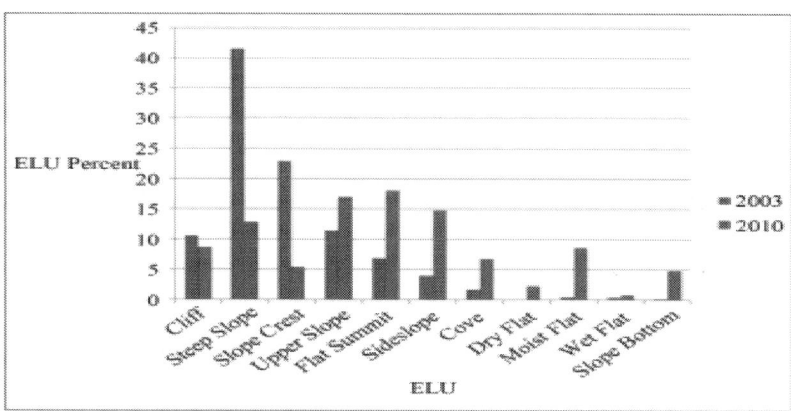

Figure 5: Comparison between landforms for areas within mines and cut or excavated between 2003 and 2010.

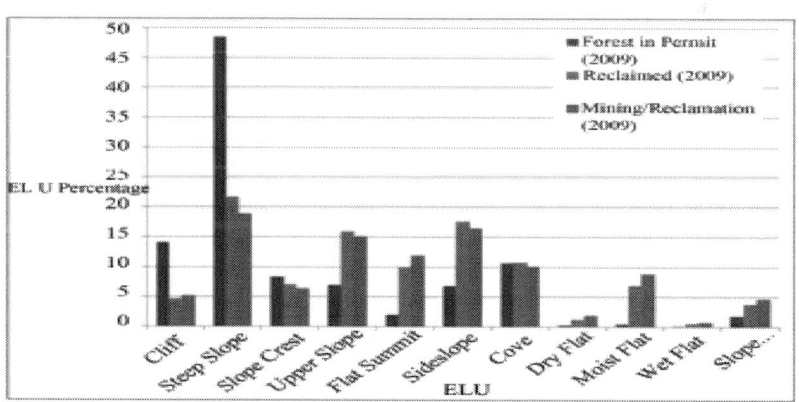

Figure 6: Comparison of landforms within different land cover categories 2009 relative to 2010 DEM.

DISCUSSION

The Southern Coalfields of West Virginia are characterized by steep slopes and narrow valleys [19]; however, this description often does not accurately describe the post-mining terrain and geomorphology. This study helped to indicate the specific landform classes that are altered in our representative study area.

The geomorphic complexity and steep slopes that characterize this region contribute to the biological richness of the Appalachian Plateau and the mixed mesophytic forest biogeographic region [27]. The Southern Coalfields of West Virginia exists within one of the most biodiverse regions within the temperate zone, and more than 2000 vascular plants exist on the landscape [28,29]. This region is globally recognized as significant for biodiversity conservation [30]. The variety of forest types is in part a result of the elevation changes and steep topography of this landscape [27].

There has been a shift in focus from rare or endangered species management to ecosystem and landscapescale management, which requires that a diversity of ecological processes be considered [31]. The methodology of this study provides an opportunity to understand landscape and terrain alterations and the ability now to examine the critical links of landforms to terrestrial habitat, ecosystem health, and biodiversity.

The question arises as to how this terrain alteration will impact biodiversity and ecosystem health. If forests are reestablished on these topographically altered mountaintops, will pre-mining forest communities and ecosystems be reestablished? Although traditional reclamation often results in grasslands or scrub/shrub lands on the reclaimed and topographically altered mountaintops [6,32] has suggested a forest reclamation approach that includes the creation of soils to support a forest community, loose grading to avoid soil compaction and increased bulk density, use of less competitive ground cover to allow for forest growth, planting of a wider variety of trees and native species, and use of proper tree planting techniques to aid in reforestation of mine scarred lands. This technique is being more commonly implemented to reclaim mine disturbed areas [32]. If such methodology is successful at creating stable and mature forest communities, will these forests represent pre-mining conditions or serve the same ecological function if the topography has been greatly altered? This question is yet to be fully explored. Assuming that topographic factors, such as slope and landscape position, are correlated with forest community type as described by [33], we suggest that it is reasonable to assume that topographic changes will result in forest community changes, and forest community variety and distribution are related to biodiversity as suggested by [27]. Terrain alteration should be considered along with the impact of introducing non-forest edge and disrupting the interior nature of the forest since MTR/VF has been shown to have the potential to alter community faunal composition [34].

SUMMARY

This study identified specific landform classes that were impacted due to cut/excavation or filled material at surface mined sites in a West Virginia, USA watershed. The landform analysis methodology provided a means to quantify and spatially assess terrain alterations using categorical landscape data derived from DEMs, a potential first step for assessing the impact of MTR/VF on terrestrial habitat and ecosystems at the landscape scale. Research is needed to understand the relationship between terrain and forest communities in the Appalachian region at both the local and regional scales. A more complete understanding of these relationships may aid in an understanding of the impact of MTR/VF on forest communities, terrestrial habitat, ecosystem health, and biodiversity.

ACKNOWLEDGEMENTS

This study was sponsored by the Appalachian Research Initiative for Environmental Science (ARIES). ARIES is an industrial affiliates program at Virginia Tech, Blacksburg, VA, USA, supported by members that include companies in the energy sector. The research under ARIES is conducted by independent researchers in accordance with the policies on scientific integrity of their institutions. The views, opinions and recommendations expressed herein are solely those of the authors and do not imply any endorsement by ARIES employees, other ARIES-affiliated researchers or industrial members. Information about ARIES can be found at http://www.energy.vt.edu/ARIES.

We would also like to acknowledge funding support provided by US Environmental Protection Agency and the West Virginia Experiment Station. This work was completed with the assistance of Adam Riley, Elise Austin, and Charles Yuill who are Remote Sensing Analyst, GIS Analyst and Landscape Analyst respectively at West Virginia University.

Lastly, we thank the anonymous reviewers who helped to improve the manuscript.

REFERENCES

1. Hooke, R.L. (1994) On the efficacy of humans as geomorphic agents. GSA Today, 4, 224-225.

2. Saylor, K.L. (2008) Land cover trends project: Central Appalachians. US Department of the Interior, Washington. http://landcovertrends.usgs.gov/east/eco69Report.html

3. Drummond, M.A. and Loveland T.R. (2010) Land-use pressure and a transition to forest-cover loss in the eastern United States. BioScience, 60, 286-298.doi:10.1525/bio.2010.60.4.7

4. Townsend, P.A., Helmers, D.P., Kingdon, C.C., McNeil, B.E., de Beurs, K.M. and Eshleman, K.N. (2009) Changes in the extent of surface mining and reclamation in the Central Appalachians detected using a 1976-2006. Landsat time series. Remote Sensing of Environment, 113, 62- 72. doi:10.1016/j.rse.2008.08.012

5. Palmer, M.A., Bernhardt, E.S., Schlesinger, W.H., Eshleman, K.N., Fourfoula-Georgiou, E., Hendryx, M.S., Lemly, A.D., Likens, G.E., Loucks, O.L., Power, M.E., White, P.S. and Wilcock, P.R. (2010) Mountaintop mining consequences. Science, 327, 148-149.doi:10.1126/science.1180543

6. Simmons, J.A., Currie, W.S., Eshleman, K.N., Kuers, K., Monteleone, S., Negley, T.L., Pohlad, B.R. and Thomas, C.L. (2008) Forest to reclaimed mine land use change leads to altered ecosystem structure and function. Ecological Applications, 18, 104-118.doi:10.1890/07-1117.1

7. Wickham, J.D., Ritters, K.H., Wade, T.G., Coan M. and Homer, C. (2007) The effect of Appalachian mountaintop mining on interior

forest. Landscape Ecology, 22, 179- 187.doi:10.1007/s10980-006-9040-z

8. Bernhardt, E.S. and Palmer, M.A. (2011) The environmental costs of mountaintop mining valley fill operations for aquatic ecosystems of the Central Appalachians. Annals of the New York Academy of Sciences, 1223, 39-57.

9. USEPA (2005) Mountaintop mining/valley fills in Appalachia. Final programmatic environmental impact statement. Region 3, US Environmental Protection Agency, Philadelphia. http://www.epa.gov/region3/mtntop/index.htm

10. Fox, J.F. and Campbell, J.E. (2009) Terrestrial carbon disturbance from mountaintop mining increases lifecycle emissions for clean coal. Environmental Science and Technology, 44, 2144-2149. doi:10.1021/es903301j

11. Hooke, R.L. (1999) Spatial distribution of human geomorphic activity in the Unites States: Comparison with rivers. Earth Surface Processes and Landforms, 24, 687- 692.doi:10.1002/(SICI)1096-9837(199908)24:8<687::AID-ESP991>3.0.CO;2-#

12. Fritz, K.M., Fulton, S., Johnson, B.R., Barton, C.D., Jack, J.D., Word, D.A. and Burke, R.A. (2010) Structural and functional characterisitcs of natural and constructed channels draining a reclaimed mountaintop removal and valley fill coal mine. Journal of the North American Benthological Society, 29, 673-689. doi:10.1899/09-060.1

13. Larson, C.F., Motyka, R.J., Arendt, A.A., Echelmeyer, K.A. and Geissler, P.E. (2007) Glacier changes in southeast Alaska and northwest British Columbia and contribution to sea level rise. Journal of Geophysical Research, 112, F01007.

14. Woolard, J.W. and Colby, J.D. (2002) Spatial characterization, resolution, and volumetric change of coastal dunes using airborne LiDAR: Cape Hatteras, North Carolina. Geomorphology, 48, 269-287. doi:10.1016/S0169-555X(02)00185-X

15. Gesch, D.B. (2005) Analysis of multi-temporal geospatial data sets to assess the landscape effects of surface mining. National Meeting of the American Society of Mining and Reclamation, ASMR, Lexington, 415-432.

16. Shank, M. (2004) Development of a mining fill inventtory from multi-date elevation data. Advanced Integration of Geospatial

Technologies in Mining and Reclamation, Surface Mining Reclamation and Enforcement, Technical Innovation and Professional Services, Denver, 1-16. http://gis.dep.wv.gov/tagis/projects/valley_fill_paper.pdf

17. Anderson, M.G., Merrill, M.D. and Biasi, F.B. (1998) Connecticut River Watershed analysis: Ecological communities and neo-tropical migratory birds. Eastern Conservation Science. The Nature Conservancy, 36.

18. WVGES (West Virginia Geologic and Economic Survey) (2005) Physiographic provinces of West Virginia. http://www.wvgs. wvnet.edu/www/maps/pprovinces.htm

19. Strausbaugh, P.D. and Core, E.L. (1977) Flora of West Virginia. Seneca Books Inc. Morgantown, 1079.

20. Grayson, R.B., Moore, I.D. and McMahon, T.A. (1992) Physically based hydrologic modeling: 1. A terrain-based model for investigative purposes. Water Resources Research, 28, 2639-2658. doi:10.1029/92WR01258

21. Environmental Systems Research Institute (ESRI) (2010) ArcGIS ArcMap Version 10.0. Redlands.

22. Moore, D.M., Lees, B.G. and Davey, S.M. (1991) A new method for predicting vegetation distributions using decision tree analysis in a geographic information system. Environmental Management, 15, 59-71. doi:10.1007/BF02393838

23. Boer, M., Del Barrio, G. and Puigdefabregas, J. (1996) Mapping soil depth classes in dry Mediterranean areas using terrain attributes derived from digital elevation models. Geoderma, 72, 99-118. doi:10.1016/0016-7061(96)00024-9

24. O'Loughlin, E.M. (1986) Prediction of surface saturation zones in natural catchments by topographic analysis. Water Resources Research, 22, 794-804.doi:10.1029/WR022i005p00794

25. Parker, A.J. (1982) The topographic relative moisture index: An approach to soil-moisture assessment in mountain terrain, Physical Geography, 3, 160-168.

26. Fels, J. and Zobel, R. (1995) Landscape position and classified landtype mapping for the statewide DRASTIC mapping project. North Carolina State University, Raleigh, 1-8.

27. Hinkle, C.R., McComb, W.C., Safely, J.M. Jr. and Schmalzer, P.A. (1993) Mixed mesophytic forests. In: Martin, W.H., Boyce, S.C. and Echternacth, A.C., Eds., Wiley, New York, 203-253.

28. Master, L.L., Flack, S.R. and Stein, B.A. (1998) Rivers of life: Critical watersheds for protecting freshwater biodiversity. Then Nature Conservancy, Arlington.

29. Stein, B.A., Kutner, L.S., and Adams, J.S. (2000) Precious heritage: The status of biodiversity in the united states. Oxford University Press, New York.

30. Ricketts, T.H., Dinerstein, E., Olson, D.M., Loucks C.J., Eichbaum, W., DellaSala, D., Kavanagh, K., Hedao, P., Hurley, P.T., Carney, K.M., Abell, R. and Walters, S. (1999) Terrestrial ecoregions of North America: A conservation perspective. Island Press, Washington.

31. Poiani, K.A., Richter, B.D., Anderson, M.G. and Richter, H.E. (2000) Biodiversity conservation at multiple scales: Functional sites, landscapes, and networks. American Institute of Biological Sciences, 50, 133-146.

32. Zipper, C.A., Burger, J.A., Skousen, J.G., Angle, P.N., Barton, C.D., Davis, V. and Franklin, J.A. (2011) Resorting forests and associated goods and services on Appalachian coal surface mines. Environmental Management, 47, 751-765. doi:10.1007/s00267-011-9670-z

33. Whittaker, R.H. (1956) Vegetation of the great smoky mountains. Ecological Monographs, 26, 1-80. doi:10.2307/1943577

34. USEPA (US Environmental Protection Agency) (2003) Draft programmatic environmental impact statement on mountaintop mining/valley fills in Appalachia.http://www.epa.gov/region3/mtntop/eis2003.htm

The Cyclic Behavior of Mountain Gravity Waves Generated by Flow over Topography*

Ziliang Li, Changji Chen, and Jinqing Liu

Department of Marine Meteorology, Physical Oceanography Laboratory, Ocean University of China, Qingdao, China

ABSTRACT

The cyclic behavior of lee wave systems, generated by stratified flow over mountains is investigated by the Advanced Regional Prediction

System (ARPS) model. The results show that, surface friction has a direct impact upon the number and timing of mountain gravity waves cycle generation. Cyclic generation of mountain lee waves and down-slope winds was found to be extremely sensitive to the magnitude of the surface drag coefficient, where mountain waves amplitude and intensity varies with the magnitude of the drag coefficient, and the interaction of mountain waves and boundary layer process determinates the wave characteristics. For the typical drag $C_d = 10^{-3}$, surface friction promotes the formation of the stationary mountain lee waves and hydraulic jump, especially, promotes boundary layer separation, the generation of low-level turbulent zones and rotor circulation or reversal flow within boundary layer. When drag coefficient becomes $C_d = 10^{-4}$, lee waves remain steady states and the first evolution cycle maintains much longer than that of $C_d = 10^{-3}$. In the case of the highest drag coefficient $C_d = 10^{-2}$, surface friction suppresses wave breaking and the onset of hydraulic jump, and reduces greatly the amplitude and intensity of lee waves and down slope wind.

INTRODUCTION

Although mountain lee waves have been studied extensively from theoretical and numerical modeling studies through to laboratory and observational experiments [1-5], the influence of surface friction effects on these waves has received little attention until recently, due partly to the difficult of introducing surface friction in theoretical models. Recent measurements and numerical simulation results suggest that processes that have received relatively little emphasis previously, such as surface friction, may be important for lee wave generation and development [6-10]. Richard [11] suggested that simulations with friction were qualitatively more realistic than simulations without surface friction, that the genesis of strong winds on the lee slope was delayed when surface friction was included, and that the downstream propagation of a mid-troposphere jump-like feature was impeded in the absence of friction. Olafsson and Bougeault [12, 13] showed that surface friction plays an active role in suppressing wave breaking for the negative impact of surface friction on wave activity. Doyle and Durran [14] demonstrate that the realistic rotors appear to develop only in the presence of surface friction by using numeric simulations with free slip and no slip lower boundary layer conditions. Peng and Thompson [15] investigated the effect of boundary layer including surface friction on flow over mountains, and they suggested that the reduction in mountain wave amplitude in the presence of surface friction is due to the reduction in the slope of the boundary layer height as compared to the terrain height. A wide body of studies has described the atmospheric gravity wave dynamics, which play a significant role in energy transport and hazard weather forecast [16-18].

These analysis, in general, conclude that surface friction reduces or suppresses wave breaking, and reduces the amplitude and intensity of lee waves, however, surface friction promotes the boundary layer separation, the generation of low-level turbulent zones and rotor circulation or reversal flow. On the other hand, meteorological observations and experimental data seem to suggest otherwise. It is thus of interest to study the effect of surface friction on flow over topography. In the present study, we seek to identify more specifically a mechanism that may provide an explanation for the effect of surface friction on lee waves. The numerical model used here is the Advanced Regional Prediction System (ARPS), described in Section 2. Section 3 is devoted to the numeric simulations of flow over topography, exploring the interaction of the flow with topography including friction dissipation. Our conclusions are presented in Section 4.

MODEL DESCRIPTION

The simulation experiments are conducted in there using the Version 5.0.0.0IHOP5 of the Advanced Regional Prediction System (ARPS), ARPS is a three-dimensional, non-hydrostatic model developed for storm scale numerical weather prediction in the University of Oklahoma. For the simulations discussed herein, considering the Coriolis force is set to zero and the atmosphere is assumed to be no surface sensible or latent heat fluxes exist, and other physical processes, such as the radiation model and the soil model are not included for the sake of simplicity.

The computational domain consists of 263 grid points with 1 km horizontal grid spacing in the horizontal direction and 83 layers in the vertical with the Rayleigh damping applied over the top half depth. The vertical grid spacing varies smoothly from 0.05 km at the ground stretching to 0.25 km near the top of the domain via

$$\begin{cases} \Delta z\,(i) = \Delta z_m + \dfrac{\Delta z_{\min} - \Delta z_m}{\tanh(2\alpha)}\tanh\left[\dfrac{2\alpha}{1-a}(i-a)\right] \\[2ex] a = \dfrac{1+(nz-3)}{2} \text{ for } i = 1, 2, \ldots, nz-3 \end{cases}$$

(1)

where Dz_m is the average grid spacing 0.25 km, α is a tuning parameter set to unity, and nz is the number of the grid points in the vertical. In addition, radiation boundary condition is applied in east-west direction, while periodic boundaries are used in north-south direction, and the big time step and small time step are 2 s and 0.5 s, respectively.

Surface friction is included in the ARPS surface physics package, which includes parameterization of surface fluxes as momentum stresses at the lower boundary:

$$-\tau_{13}\big|_{surface} = \overline{\rho}C_d\left[|V|u - |\overline{V}|\overline{u}\right]$$

(2)

$$-\tau_{23}\big|_{surface} = \overline{\rho}C_d\left[|V|v - |\overline{V}|\overline{v}\right]$$

(3)

where t_{13} is the momentum stress on the constant x plane and t_{23} is the momentum stress on a constant y plane in the vertical direction, C_d is the non-dimensional drag coefficient, $|V|$ is wind speed, and over-bars represent the base state. And the parameterization of the vertical turbulent diffusion follows the 1.5-order turbulent kineticenergy closure.

The basic state velocity $U = 20$ ms^{-1} and static stability $N^2 = 0.01$ s^{-1}, and the mountain profile used in here is the bell-shaped mountains, given as

$$h(x, y) = \dfrac{h_{\max}a^2}{1+(x-x_0)^2}$$

(4)

where h_{max} is the maximum height of mountain, the half width of mountain a = 10 km, and the mountain center location in horizontal direction x_0 = 72 km.

RESULTS AND INTERPRETATIONS

We first consider the atmosphere to be uniform with height far upstream of the mountain to identify the evolution of lee wave systems. In the absence of surface friction, vertically propagating gravity waves typically form when the maximum height of mountain h_{max} = 0.5 km. However, when the maximum height of mountain increases to h_{max} = 2 km, then the airflow may be blocked on the upwind slope and the trapped lee waves form on the downstream lee side, accompanied by the hydraulic jumps.

When surface friction is activated, the flow pattern in the lower atmosphere becomes quite different from that of free slip case. When choosing h_{max} = 0.5 km, it is extremely sensitive to the distance Z_h between the surface (u = v = w = 0) and the first model grid point layer. Such as Z_h = 25 m, the influence of surface friction on the vertical propagating waves is relatively minor compared with the free-slip case. However, if Z_h increases from 25 m to 125 m, due to the deeper depth of the positive vertical wind shear in the boundary layer, surface friction is known to reduce the tendency toward wave breaking, apart from the general reduction of the near surface winds speed, these results are consistent with previous work [6,11,12,15], especially under no slip conditions.

However, if choosing h_{max} = 2 km, surface friction has a direct impact upon the number and timing of lee wave cycles. In the first cycle of wave evolutions, the results are consistent with the simulations of Richard [11] and others work, but involved in cyclic generation of lee wave systems in our run, the results are different from others for the regeneration of lee waves in the second cycle. Most significantly, the effect of surface friction on lee wave amplitude and intensity varies with the magnitude of drag coefficient, especially, cyclic generation of lee waves and down slope winds in our runs was found to be extremely sensitive to the magnitude of drag coefficient. In addition, surface friction promotes the boundary layer separation and the forming of rotors within the boundary layer.

For the typically surface friction drag coefficient $C_d = 10^{-3}$, the cyclic regeneration of lee wave systems is apparent and allows the wave systems to undergo one transition of two cycles in the whole integration of 144,000 sin which lee wave system begin to dissipate and complete one cycle after the integration of almost 26,000 s and they begin to regenerate and intensify again after integration of almost 36,000 s, the interaction of lee waves and rotors circulation finally determine the flow characters. Firstly, we can see that from Figures 1 and 2, in the first cycle stage, lee waves intensify and low level closed rotor circulation form after integration of almost 12,000 s, where the low level rotors and rotor circulation play a negative role in the amplitude and intensity of lee waves, and lee waves vanish for surface friction after integrating of almost 26,000 s. In the second cycle stage, lee waves begin to occur in the middle and upper atmosphere after integrating of 36,000 s. As time increases, then hydraulic jump, large amplitude lee waves, lee vortex, organized rotor and rotor circulation regenerate after integrating of 76,000 s. Secondly, surface friction reduces the magnitude of the down slope wind, suppresses the amplitude and intensity of lee waves in the first cycle. However, in the second cycle, the influences of surface friction upon the magnitude of the down slope wind appears to be relatively minor, and surface frictions have a minimal impact upon the structures of lee waves in the middle and upper above the boundary layer. Such as, the maximum cross ridge speed attains 80.2 ms^{-1} in free-slip condition, but for the drag coefficient $C_d = 10^{-3}$, it almost attains the value 57.8 ms^{-1} in the first cycle and almost the same value 78 ms-1 in the second cycle, respectively.

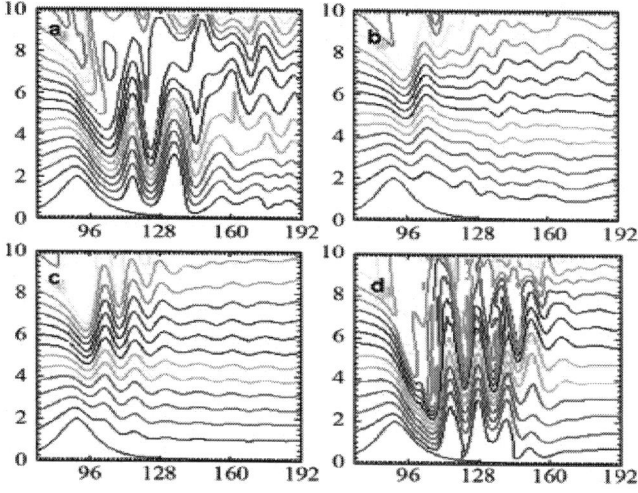

Figure 1: Potential temperature (K) for the simulation of flow over a 2 km mountain with surface friction of $C_d = 10^{-3}$, after integration of (a) 12,000 s; (b) 26,000 s; (c) 36,000 s; (d) 76,000 s.

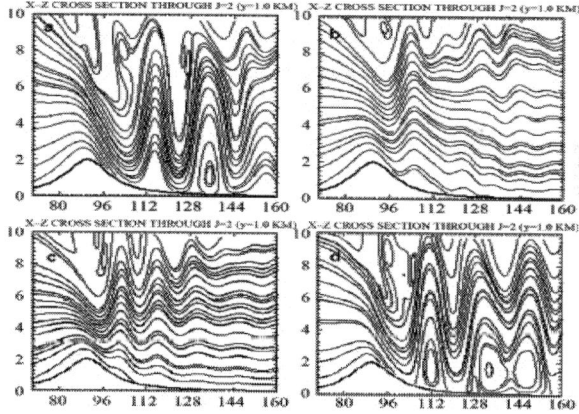

Figure 2: Streamlines for the simulation of flow over a 2 km mountain with surface friction of $C_d = 10^{-3}$, after integration of (a) 12,000 s; (b) 26,000 s; (c) 36,000 s; (d) 76,000 s.

When the drag coefficient is smaller than the typically value $C_d = 10^{-3}$, that is, the drag coefficient reduced by $C_d = 10^{-4}$, except low level rotors and rotor circulations are established in the first cycle, compared with the freeslip case, the influences of surface friction upon the magnitude of lee waves and down slope wind appears to be relatively minor, but the regeneration of lee waves is delayed greatly compared with the case $C_d = 10^{-3}$. The potential temperature, streamline and horizontal velocity given by Figures 3 and 4 show that, in the first cycle, hydraulic jump lee waves and lee vortex occur above the boundary layer with lee rotors within the boundary layer, and the maximum cross ridge speed is attains 82.2 ms^{-1} in the first cycle. However, it is quite different from the free-slip case and the case $C_d = 10^{-3}$, lee wave systems have dissipated for surface friction after integrating of 36,000 s, and they will begin to regenerate slowly again after integrating of 126,000 s, the time interval between the two cycles is apparent elongated, and the life span of the first cycle increases nearly 18,000 s and the second cycle is delayed nearly 76,000 s compared with the case $C_d = 10^{-3}$. This sensitivity of lee wave to drag coefficient suggests that there is an optimum value of surface friction (approximately $C_d = 10^{-4}$) that maximizes the strength of the mountain wave systems for this particular simulation.

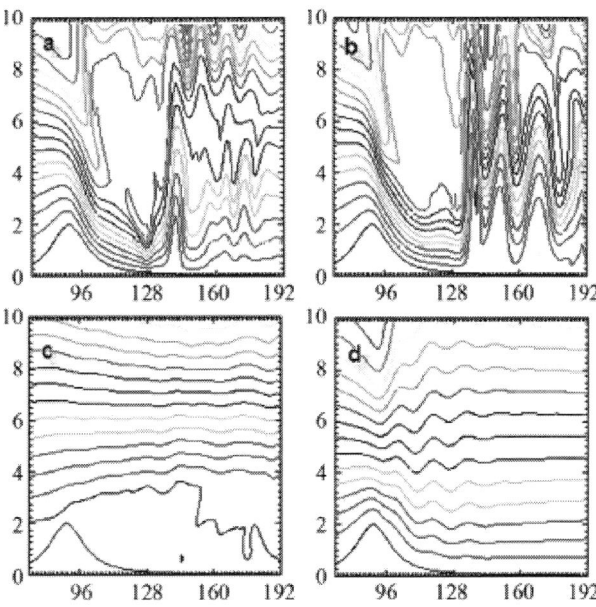

Figure 3: Same as Figure 2 but for the case with surface friction of $C_d = 10^{-4}$.

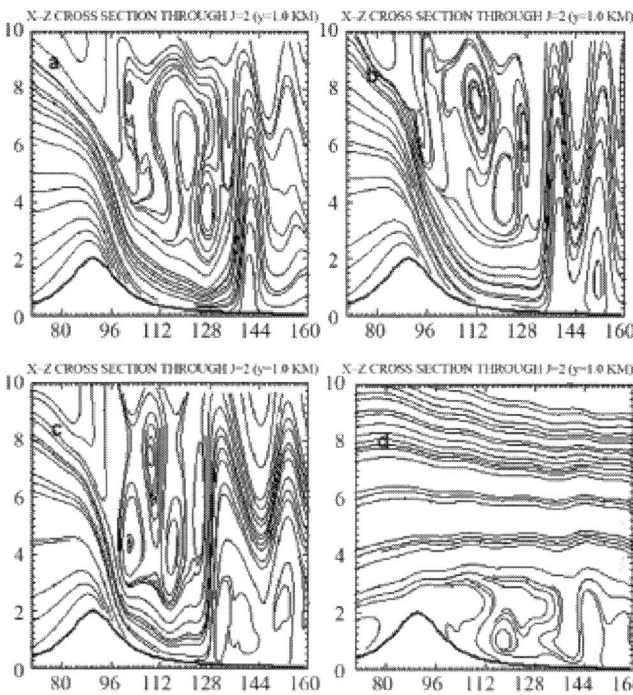

Figure 4: Same as Figure 2 but for the case with surface friction of $C_d = 10^{-2}$.

When surface friction drag coefficient is increased, for the case of $C_d = 10^{-2}$, surface friction suppresses wave breaking and the onset of hydraulic jump, and reduces greatly the amplitude and intensity of lee waves and down slope wind, wherein the lee waves show the small amplitude of non-hydraulic jump type with the maximum cross-ridge wind only 20 ms^{-1} in the first cycle and 40 ms–1 in the second cycle. The potential temperature, streamline and horizontal velocity given by Figures 5 and 6 show that, in the first cycle stage, the duration of wave systems are shortened greatly and lee waves have dissipated after integrating of 12,000 s. In the second cycle stage, the small amplitude lee waves regenerate above the boundary layer after integrating of 36,000 s, as time increases, the amplitude and intensity of lee waves of the boundary layer aloft increase greatly, accompanied by the rotors and reversal flow in the boundary layer after integrating of 76,000 s. Finally, the lee waves of nonhydraulic jump type have dissipated after integrating of 126,000 s and complete the second cycle stage.

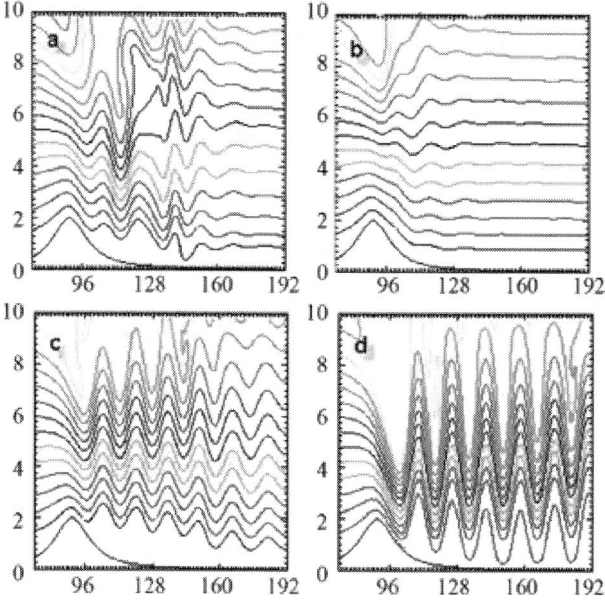

Figure 5: Same as Figure 1 but for the case with surface friction of $C_d = 10^{-4}$.

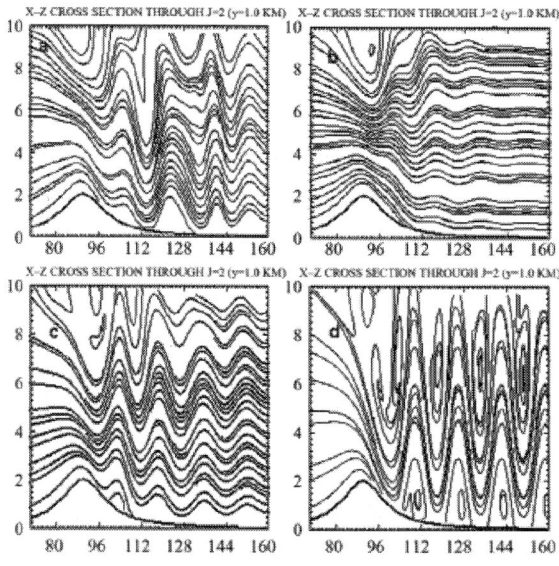

Figure 6: Same as Figure 2 but for the case with surface friction of $C_d = 10^{-2}$.

DISCUSSION AND CONCLUSIONS

The stratified flow over mountain is investigated by the Advanced Regional Prediction System (ARPS) model, the results show that the regeneration of lee wave systems is apparent and allows the wave systems to undergo one transition of two cycles. Most significantly, the effect of surface friction on lee wave amplitude and intensity varies with the magnitude of the drag coefficient, especially, cyclic regeneration of lee waves and down slope winds in our runs was found to be extremely sensitive to the magnitude of the drag coefficient. For the typical drag values $C_d = 10^{-3}$, surface friction promotes the formation of the stationary mountain lee waves and the stationary hydraulic jump, especially, promotes the boundary layer separation, the generation of the low-level turbulent zones and rotor circulation or reversal flow within the boundary layer. When the drag coefficient is smaller $C_d = 10^{-4}$, the lee waves remain steady states and the first cycle maintains much longer than that of the case $C_d = 10^{-3}$. In the case of the highest drag coefficient ($C_d = 10^{-2}$), surface friction suppresses wave breaking and the onset of hydraulic jump, and reduces greatly the amplitude and intensity of lee waves and down slope wind. It can be concluded that, the interaction of lee waves and the boundary layer process, lee waves induce rotors and rotors induce lee waves, finally determine the characters of lee wave systems.

Extending to the case of three-layer flow over mountain, that is, the static stability of the lower (below 2.5 km), middle layer (2.5 km - 5 km) and upper atmosphere (above 5 km) are 0.02, 0.01 and 0.015, respectively, and keep the same physical domain and the other parameters as before. The results show that the train of lee waves becomes the dominant feature, and lee waves and rotors show the same characters of cyclic regeneration as the uniformly stratified flow over mountain. However, extending to the mean state critical layer flow over mountain, the base state wind is 20 ms^{-1} below 5 km and is reduced from 20 ms^{-1} at 5 km to zero above 7 km. It suggests that the critical layer increases the amplitude and intensity of lee waves and surface friction promotes the forming of lee rotors, but the evolution of lee waves is relatively minor as time increases both with and without surface friction. The responses of flow to topography are mainly determined by the factors of the topography, stratification and wind configuration.

REFERENCES

1. R. B. Smith, "The Influence of Mountains on the Atmosphere. Advances in Geophysics," Academic Press, Amsterdam, 1979, pp. 87-230. doi:10.1016/S0065-2687(08)60262-9

2. R. B. Smith, "Stratified Flow over Topography. Environment Stratified Flows," Academic Press, Amsterdam, 2002, pp. 121-162.

3. D. R. Durran, "Mountain Waves and Down-Slope Flows. Atmospheric Processes over Complex Terrain, Metero. Monographs," American Meteorological Society, Vol. 23, No. 45, 1990, pp. 59-81.

4. P. G. Baines, "Topographic Effects in Stratified Flows," Cambridge University Press, Cambridge, 1995, pp. 1-488.

5. M. G. Wurtele, R. D. Sharman and A. Data, "Atmospheric Lee Waves," Annual Review of Fluid Mechanics, Vol. 28, 1996, pp. 429-476. doi:10.1146/annurev.fl.28.010196.002241

6. V. Grubisic and P. K. Smolarkiewicz, "The Effect of Bottom Friction on Shallow-Water Flow Past an Isolated Obstacle," Journal of the Atmospheric Sciences, Vol. 54, No. 11, 1997, pp. 1943-1960.

7. G. R. Mamatsashvili, V. R. Avsarkisov, G. D. Chagelishvili, R. G. Chanishvili and M. V. Kalashnik, "Transient Dynamics of Nonsymmetric Perturbations Versus Symmetric Instability in Baroclinic Zonal Shear Flows," Journal of the Atmospheric Sciences, Vol. 67, No. 9, 2010, pp. 2972-2989. doi:10.1175/2010JAS3313.1

8. M. E. McIntyre, "Spontaneous Imbalance and Hybrid Vortex Gravity Structures," Journal of the Atmospheric Sciences, Vol. 66, No. 5, 2009, pp. 1315-1326.doi:10.1175/2008JAS2538.1

9. R. Plougonven, C. Snyder and F. Zhang, "Comments on Application of the Lighthill-Ford Theory of Spontaneous Imbalance to Clear-Air Turbulence Forecasting," Journal of the Atmospheric Sciences, Vol. 66, No. 8, 2009, pp. 2506-2510. doi:10.1175/2009JAS3027.1

10. C. Snyder, R. Plougonven and D. J. Muraki, "Mechanisms for Spontaneous Gravity Wave Generation within a Dipole Vortex,"

Journal of the Atmospheric Sciences, Vol. 66, No. 11, 2009, pp. 3464-3478. doi:10.1175/2009JAS3147.1

11. E. Richard, R. Mascart and E. C. Nicherson, "The Role of Surface Friction in Down Slope Windstorms," Journal of Applied Meteorology, Vol. 28, No. 4, 1989, pp. 241-251. doi:10.1175/1520-0450(1989)028<0241:TROSFI>2.0.CO;2

12. H. Olafsson and P. Bougeault, "The Effect of Rotation and Surface Friction on Orographic Drag," Journal of the Atmospheric Sciences, Vol. 54, No. 1, 1997, pp. 193-210.doi:10.1175/1520-0469(1997)054<0193:TEORAS>2.0.CO;2

13. H. Olafsson and P. Bougeault, "Why Was There No Wave Breaking in PYREX?" Beitraege zur Physik der Atmosphaere, Vol. 70, No. 2, 1997, pp. 167-170.

14. J. D. Doyle and D. R. Durran, "The Dynamics of Mountain-Wave Induced Rotors," Journal of the Atmospheric Sciences, Vol. 59, No. 2, 2002, pp. 186-201. doi:10.1175/1520-0469(2002)059<0186:TDOMWI>2.0.CO;2

15. M. S. Peng and W. T. Thompson, "Some Aspects of the Effects of Surface Friction on Flow over Mountains," Meteorological Society, Vol. 129, No. 593, 2003, pp. 2527- 2557.doi:10.1256/qj.02.06

16. H. M. Albert, A. Joan and P. Riwal, "On the Intermittency of Gravity Wave Momentum Flux in the Stratosphere," Journal of the Atmospheric Sciences, Vol. 69, No. 11, 2012, pp. 3433-3448. doi:10.1175/JAS-D-12-09.1

17. J. Knox, D. McCann and P. Williams, "Application of the Lighthill-Ford Theory of Spontaneous Imbalance to ClearAir Turbulence Forecasting," Journal of the Atmospheric Sciences, Vol. 65, No. 10, 2008, pp. 3292-3304. doi:10.1175/2008JAS2477.1

18. F. Lott and R. Plougonven, "Gravity Waves Generated by Sheared Three-Dimensional Potential Vorticity Anomalies," Journal of the Atmospheric Sciences, Vol. 69, No. 7, 2012, pp. 2134-2151. doi:10.1175/JAS-D-11-0296.1

Estimation of Potentially Toxic Elements Contamination in Anthropogenic Soils on a Brown Coal Mining Dumpsite by Reflectance Spectroscopy: A Case Study

Asa Gholizadeh[1], Luboš Borůvka[1], Radim Vašát[1], Mohammadmehdi Saberioon[2], Aleš Klement[1], Josef Kratina[1], Václav Tejnecký[1], and Ondřej Drábek[1]

[1]Department of Soil Science and Soil Protection, Faculty of Agrobiology, Food and Natural Resources, Czech University of Life Science, Kamýcká 129, 165 21- Suchdol, Praha 6- Prague, Czech Republic

[2]Laboratory of Image and Signal Processing, Institute of Complex Systems, Faculty of Fisheries and Protection of Waters, University of South Bohemia in České Budějovice, Zámek 136 37 333- Nové Hrady, Czech Republic

ABSTRACT

In order to monitor Potentially Toxic Elements (PTEs) in anthropogenic soils on brown coal mining dumpsites, a large number of samples and cumbersome, time-consuming laboratory measurements are required. Due to its rapidity, convenience and accuracy, reflectance spectroscopy within the Visible-Near Infrared (Vis-NIR) region has been used to predict soil constituents. This study evaluated the suitability of Vis-NIR (350–2500 nm) reflectance spectroscopy for predicting PTEs concentration, using samples collected on large brown coal mining dumpsites in the Czech Republic. Partial Least Square Regression (PLSR) and Support Vector Machine Regression (SVMR) with cross-validation were used to relate PTEs data to the reflectance spectral data by applying different preprocessing strategies. According to the criteria of minimal Root Mean Square Error of Prediction of Cross Validation ($RMSEP_{cv}$) and maximal coefficient of determination (R^2_{cv}) and Residual Prediction Deviation (RPD), the SVMR models with the first derivative pretreatment provided the most accurate prediction for As (R^2_{cv} = 0.89, $RMSEP_{cv}$ = 1.89, RPD = 2.63). Less accurate, but acceptable prediction for screening purposes for Cd and Cu ($0.66 < R^2_{cv} < 0.81$, $RMSEP_{cv}$ = 0.0.8 and 4.08 respectively, $2.0 < RPD < 2.5$) were obtained. The PLSR model for predicting Mn (R^2_{cv} = 0.44, $RMSEP_{cv}$ = 116.43, RPD = 1.45) presented an inadequate model. Overall, SVMR models for the Vis-NIR spectra could be used indirectly for an accurate assessment of PTEs' concentrations.

INTRODUCTION

Our society and civilization now rely heavily on the mining industry to sustain our way of living. However, mining is one of the anthropogenic activities that causes some of the most dramatic disturbances to the earth. Fertile, cultivated land is transformed into wasteland, as mining activities generate a vast amount of solid wastes, which are deposited at the surface and typically occupy a huge area [1, 2]. Among the various geo-environmental impacts of mining, contamination of soil is by far the most significant. Elevated concentrations of Potential Toxic Elements (PTEs) in soils do not only impact the soil quality, but due to their persistent nature and long biological half-lives, can accumulate in the food chain and can eventually influence human health [3–5]. Although the adverse effects of PTEs have long been known, and exposure to PTEs continues (and is even increasing in some areas), most of the former metallurgical tailing dumpsites are now abandoned. However, no particular safety measures are in place, and their environmental impact has received little attention. Potentially Toxic Elements (PTEs) concentrations in soils can be measured, but their determination depends on large-scale sampling and physical or conventional analysis techniques, which are time-consuming, inefficient, and expensive when applied on a large scale [6]. Moreover, according to Xie et al. [5], conventional methods for environmental soil monitoring require the collection of numerous samples, followed by laboratory analyses that involve complex processes such as separation and pre-concentration. In practice, sampling density and analytical diversity are frequently less than sufficient due to the significant costs of analyses.

Diffuse reflectance spectroscopy technique is a low cost technique with little or no sample preparation, and has been considered as an alternative to conventional soil analytical methods [7, 8]. It has shown to be a powerful tool for such studies in agricultural applications as it provides knowledge of the state of soil, giving results on-site in real-time. Furthermore, this method can be adjusted to provide results for more than one soil attribute of the soil with a single analysis [9]. Some researchers used Visible (Vis) and Near Infrared (NIR), ranging from 350 nm to 2500 nm, to analyze the spectrally active properties of sediment and soil samples. Toxic elements in soils can often be absorbed or bound by these spectrally active constituents [10]. This makes it possible to study the characteristics of PTEs in soils using Vis and NIR spectroscopy [3]. Kemper and Sommer [11] successfully used reflectance spectroscopy to estimate arsenic (As), iron (Fe), mercury (Hg), lead (Pb), sulfur (S) and antimony (Sb) contents in the Aznalcollar Mine area in Spain. Bray et al. [12] also used Vis, NIR and Mid-Infrared (MIR) spectroscopy, calibrated by ordinal logistic regression, for the screening of either contaminated or uncontaminated soil at different thresholds for copper (Cu), zinc (Zn), cadmium (Cd) and Pb. It is believed that Vis-NIR can substantially decrease both the time and costs associated with screening for PTEs.

Chemometric methods are often needed to analyze the spectra characteristics of soil [13]. Using a set of well-known calibration methods makes this process feasible. Choosing the most robust calibration technique can help to achieve a more reliable prediction model. Multiple Linear Regression (MLR) [14], Principle Component Regression (PCR) [15], or Partial Least Squares Regression (PLSR) [10, 16] have been used in the past to build models for estimating the content of toxic elements in soil or sediments. All of the above-mentioned calibration methods require the creation of robust and generalized models, due to their potential tendency to over-fit the data [8, 17]. Therefore, using a method that can overcome the problems of other calibration methods, such as Support Vector Machine Regression (SVMR) seems essential. According to Vapnik [17], SVMR is a supervised non-parametric statistical learning technique; thus it represents a different model class compared with the previous techniques.

To the best of our knowledge, the SVMR technique has not yet been used to analyse soil contamination, in the spectral domain. This study was conducted to assess selected PTEs, namely manganese (Mn), Cu, Cd, Zn, Fe, Pb and As concentrations in anthropogenic soils on brown coal mining dumpsites using Vis-NIR. We evaluate the feasibility of the technique for the rapid prediction of the above-mentioned contaminants, and to compare the performance of PLSR and SVMR methods for multivariate calibrations using soil reflectance spectra. It was envisaged that this rapid and inexpensive method for obtaining accurate information of PTEs would be valuable in providing reference data for soil environment monitoring by proximal and remote sensing.

MATERIALS AND METHODS

Study Area and Soil Sampling

Six dumpsites, which were located in mines Bílina and Tušimice (Fig. 1), in the Czech Republic, were selected: Pokrok (50° 60′ N; 13° 71′ E), Radovesice (50° 54′ N; 13° 83′ E), Březno (50° 39′ N; 13° 36′ E), Merkur (50° 41′ N; 13° 30′ E), Prunéřov (50° 42′ N; 13° 28′ E) and Tumerity (50° 37′ N; 13° 31′ E). Permission to enter the studied areas was issued by the mining company Severoceske Doly, a.s., Chomutov (www.sdas.cz), which manages these areas, and soil sampling was made under the supervision of its representatives. The land is protected until the reclamation process is completed, for security reasons rather than for natural protection. No protected species were collected.

All dumpsites are formed by clay. On a part of each dumpsite, a cover with natural topsoil was spread in an amount of approximately 2500 to 3000 t per ha one year before sampling. Topsoil material originated from humic horizons of natural soils of the region, particularly Vertisols, and partly also Chernozems (clayic and haplic). Topsoil was not mixed with the dumpsite material. Individual soil properties differed slightly between the six dumpsites. Some characteristics of the soils, including pH, Soil Organic Matter (SOM) and texture were measured using bulk control subsamples since they are important environmental indicators. Specifically, the soil pH range for the area was 5.3–8.5. The SOM content range was 0.6–3.8%. Texture analysis, which was performed by the hydrometer method, showed that soil of the area had 37.30% clay, 33.10% sand and 29.60% silt. Disturbed and undisturbed soil samples were randomly collected at all of the dumpsites randomly. 103 soil samples were collected on the Pokrok dumpsite, 40 samples on the Radovesice dumpsite, 25 samples on the Březno, 38 samples on the Merkur dumpsite, 48 samples on the Prunéřov dumpsite, and 10 samples on the Tumerity dumpsite. Approximately half of the sampling points were located on the area with natural topsoil cover, and on the area without the cover. Sampling was made in the depth of 0 to 20 cm. This depth corresponds to the common depth of a ploughing soil layer, as these soils will be used as arable land in future. Where it was applied, the depth of the topsoil cover was also at least 20 cm.

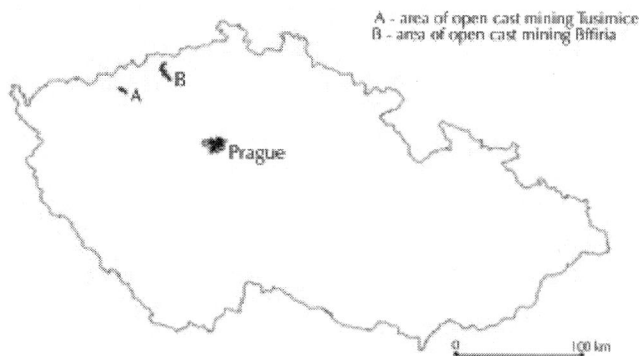

Figure 1: Map of the sampling locations in the Czech Republic.

Soil Analysis

Samples were air-dried, then sieved through a 2 mm mesh. All samples were then used for analyses of PTEs (including Cu, Mn, Cd, Zn, Fe, Pb and As) and reflectance measurements. Target elements were extracted using 2M HNO_3 [5, 18]. Arsenic was determined in extracts by a flow-through electrochemical coulometry analyser EcaFlow 150GLP (Istran, SK). All other elements were measured by atomic absorption spectrometry (Varian Spectra AA280, Varian, Australia). Samples and standards were matrix matched and all analyses were performed in triplicates.

Reflectance Spectroscopy Measurement

Reflectance was measured in the 350–2500 nm wavelength range by a FieldSpec 3 spectroradiometer (Analytical Spectral Devices Inc., USA) with a contact probe under standard laboratory conditions. The spectral resolution of the spectroradiometer was 3 nm for the region 350–1000 nm and 10 nm for the region 1000–2500 nm. Furthermore, the radiometer bandwidth from 350–1000 nm was 1.4 nm, and from 1000–2500 nm, bandwidth was 2 nm. Samples were illuminated using a stable direct current powered 50 W tungsten-quartz-halogen lamp, which was mounted on a tripod. The angle of incident illumination was 15° and the distance between the illumination source and the sample was 30 cm. A fiber-optic probe with 8° field of view was used to collect reflected light from the sample. The probe was mounted on a tripod and positioned approximately about 10 cm vertically above the sample. The sample dish was over-filled with soil sample, and then leveled off using a blade to ensure a flat surface that is flush with the top of the dish. The final spectrum was an average based on 20 iterations from 4 directions, with 5 iterations per direction to increase the signal-to-noise ratio. Each sample spectrum was corrected for background absorption by division of the reference spectrum of a standardized white $BaSO_4$ panel.

Model Construction and Validation

Outliers are commonly defined as observations that are not consistent with the majority of the data [19], such as observations that deviate significantly from normal values. An outlier can be defined as (i) a spectral outlier when the sample is spectrally different than the rest of the samples, and (ii) a concentration outlier, which occurs when the predicted value has a residual difference significantly greater than the mean of the predicted values (>2.5 times). For all samples, an exploratory analysis was carried out to detect outliers before establishing the regression model [3]. Murray [20] noted that removing outliers may increase prediction accuracy; hence the outliers were left out. Correlation between PTEs concentration and reflectance spectra was determined using Pearson's correlation.

It was necessary to calibrate a model that provides accurate predictive performance about the quantity of PTEs contained in each soil sample; the captured soil spectra, together with laboratory data of PTEs were imported into R software (R Development Core Team, 2011) to be processed. From a total of 264 samples, subsets were mostly used to determine the content of PTEs. The number of samples subjected to individual analysis was then as follows: the entire data were tested for Mn and Fe; 148 samples were tested for Pb; 115 for Cu and Zn, and 104 samples for Cd and As. Spectral preprocessing techniques are a variety of mathematical methods for correcting light scattering in reflectance measurements and data enhancement before the data was used in calibration models. Spectral derivative transformation is actually one of the best methods for removing baseline effects [21]. The first derivative is very effective for removing baseline offset; the second derivative is very effective for both the baseline offset and linear trend from a spectrum [21, 22]. In this study, prior to all further spectra treatments, the noisy part of the spectra range (350–399 nm and 2450–2500 nm) was cut out and then the spectra were subjected to Savitzky-Golay smoothing with a second-order polynomial fit and 11 smoothing points [6, 10]. This was done in order to remove the artificial noise caused by the spectroradiometer instrument. Predictive models were fitted based on smoothed raw spectra at first, and then two types of preprocessed spectra were used, as with first and second derivative manipulations which were calculated using the Savitzky-Golay algorithm. Moreover, PLSR and SVMR models were employed to calibrate spectral data with chemical reference data, and to describe the relationship between reflectance spectra and measured PTEs.

The PLSR technique is closely related to Principle Component Regression (PCR). However, unlike PCR, the PLSR algorithm selects successive orthogonal factors that increase the covariance between predictor (X, spectra) and response variables (Y, laboratory data). By fitting a PLSR model, one hopes to find a few PLSR factors that explain most of the variation in both predictors and responses [23]. Partial Least Squares Regression (PLSR) handles multicollinearity. It is robust in terms of data noise and missing values, and unlike PCR, it balances the two objectives of explanation response and predictor variation (thus calibrations and predictions are more robust) and presents the decomposition and regression in a single step [23]. PLSR models were fitted with the *pls* R package [24], using the classical orthogonal scores algorithm. The optimal number of latent variables (factors) was determined by minimizing the value of Root Mean Square Error of Prediction (RMSEP) by leave-one-out cross validation [5].

The concept of SVMR follows a different approach of supervised learning. Its algorithm is based on the statistical learning theory [25]. It has been known to strike the right balance between accuracy attained on a given finite amount of training patterns, and an ability to generalize to unseen data. The most valuable properties of SVMs are their ability to handle large input spaces efficiently, to deal with noisy patterns and multi-modal class distributions, and their restriction on only a subset of training data in order to fit a non-linear function [8, 26]. For SVMR prediction we used radial basis function kernel, as contained in *e1071* R package [27].

Because a limited number of samples were available, the validation was done using the leave-one-out cross validation procedure with computing corresponding index of determination in cross validation (R^2_{cv}) and Root Mean Squared Error of Prediction of Cross Validation ($RMSEP_{cv}$). The leave-one-out cross validation means that in turn, one sample was taken out from the data set and the calibration was made based on the remaining samples, and then the prediction was made for the sample that was taken out at the beginning. Each time, *n-1* samples were used to build the regression model from all *n* samples within the dataset [19]. The same procedure is repeated for each sample in the sample set. The differences between observed and predicted values give the R^2_{cv} and $RMSEP_{cv}$ which is a standard method for validation of the prediction models [5, 6].

Accuracy Assessment of Techniques

Assessment of prediction accuracy of the models was carried out using a leave-one-out cross-validation approach (R^2_{cv} and $RMSEP_{cv}$) [5, 28, 29, 30], the value of RPD was also calculated which is an useful indicator of the practical utility of a calibration model to predict soil property considering the variation of the soil property [5]. The $RMSEP_{cv}$ and RPD were computed as follows:

$$RMSEP_{cv} = \sqrt{\frac{1}{N} \sum_{i=1}^{N} (y'_i - y_i)^2}$$

where y'_i is the predicted and y_i is the observed value. The smallest $RMSEP_{cv}$ value was related to the optimal calibration model.

$$RPD = \frac{SD}{RMSEP_{cv}}$$

where SD is Standard Deviation.

RESULTS AND DISCUSSION

Soil Samples Descriptive Statistics

General statistical results of PTEs in the six dumpsites are summarized in Table 1.

Table 1: Descriptive statistics of PTEs in the studied sample set according to location

Item	Cu	Mn	Fe	Cd mg/kg	Pb	Zn	As
			Pokrok (n = 103)				
Min	5.50	198.3	2503.4	0.01	7.60	8.30	0.49
Max	35.70	869.1	9752.6	0.73	42.40	127.10	19
Mean	13.76	599.4	5418	0.27	18.43	25.26	4.48
Std.	3.58	118.6	1330.1	0.11	5.32	15.77	3.39
C.V. (%)	26	20	25	40	29	62	76
			Radouesice (n = 40)				
Min	6.42	254.1	1754.4	0.03	4.70	9.38	0.18
Max	22.10	844.1	6876.9	0.30	49.60	66.85	1.30
Mean	14.20	541.3	4489.3	0.17	13.70	21.98	0.67
Std.	3.45	125.1	974.4	0.05	6.40	11.15	0.25
C.V. (%)	24	23	22	30	47	51	38
			Březno (n = 25)				
Min	9.01	473.3	2398.5	0.00	10.90	11.49	0.49
Max	38.81	885.8	31281.8	0.37	21.60	200.27	5.89
Mean	14.37	630.9	9967.2	0.16	14.17	41.50	1.12
Std.	5.95	105.9	103.58.2	0.11	2.97	41.62	1.04
C.V. (%)	41	16	104	64	21	100	93
			Merkur (n = 38)				
Min	7.29	318	2361.8	0.04	9.30	6.95	0.33
Max	16.76	787.3	8047.7	0.27	55.90	32.22	9.57
Mean	12.22	590	4852.7	0.16	17.53	13.56	0.97
Std.	1.77	100.7	1355.6	0.06	7.23	4.19	1.45
C.V. (%)	14	17	28	39	41	31	149
			Prunéřov (n = 48)				
Min	8.40	41.6	2105	0.00	0.90	6.60	0.00
Max	92.24	984	9225.4	0.24	24.80	213.11	3.30
Mean	15.81	582.6	5532.5	0.11	14.36	26.83	0.98
Std.	14.36	224.4	1595.5	0.06	4.82	39.32	0.86
C.V. (%)	91	41	29	55	34	147	87
			Tumerity (n = 10)				
Min	12.29	496.6	4163.8	0.00	9.50	15.50	0.37
Max	20.34	1027.6	6484.3	0.20	14.50	48.56	0.51
Mean	15.03	753.1	6702.3	0.12	12.25	25.61	0.42
Std.	2.40	192.3	1426.6	0.05	1.38	10.32	0.05
C.V. (%)	16	26	21	44	11	40	12

doi:10.1371/journal.pone.0117457.t001

A comparison of Coefficients of Variation (C.V.) of different contaminants showed that among all parameters, As had the highest C.V., especially in the Merkur area (149%). Hence As varied the most as compared to other measured parameters. In contrast, Pb in Tumerity (11%) had the lowest C.V., which shows that its distribution is more homogenous than the other PTEs.

The estimated mean concentrations of Cd (0.27 mg/kg), Pb (18.43 mg/kg) and As (4.48 mg/kg) were higher in Pokrok than other locations, which may suggest a threat toward soil destined for agricultural soils. The mean concentration of Fe (9967.19 mg/kg) in the studied soil sample of Březno was also high, which was probably related to the iron oxide-rich characteristic of the soil type and its formation.

Vis-NIR Reflectance Spectroscopy of the Soil Samples and Data Preprocessing

Visual inspection of the spectra allowed detection of some spectral readings, possibly affected by measurement errors. These were removed, and the final spectral library had a total of 264 soil spectra. Raw reflectance, smoothed spectra by Savitzky-Golay, and first and second derivative spectra of all selected soil samples in Pokrok (the location that had the most samples) are shown in Fig. 2. Other locations also showed the same pattern.

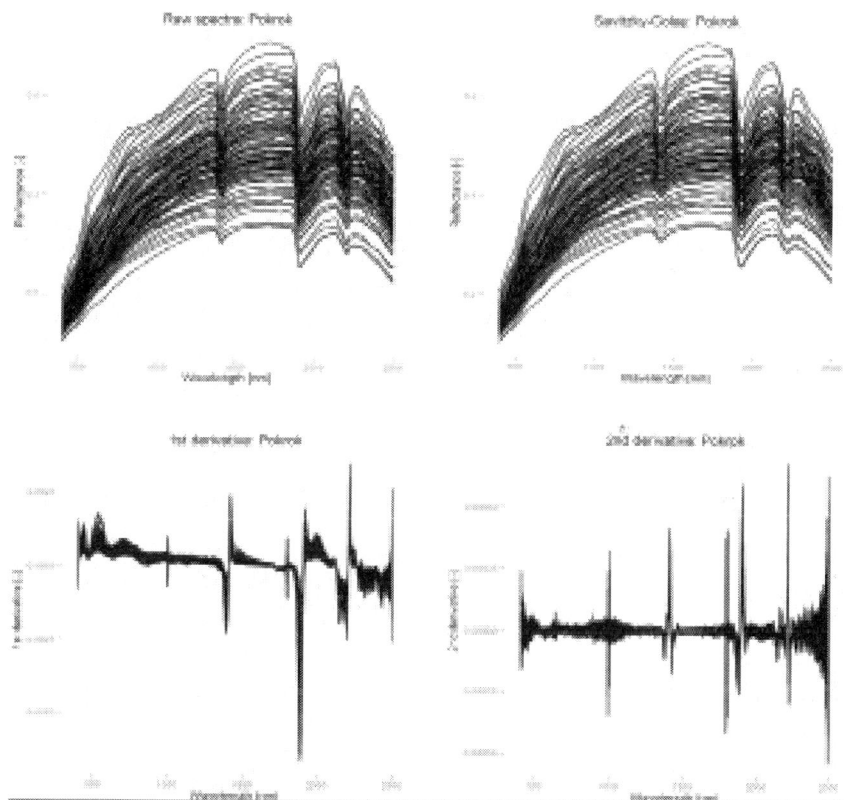

Figure 2: Raw reflectance spectra, smoothed spectra by Savizky-Golay and preprocessed spectra of soil samples for Pokrok.

As can be seen in Fig. 2 (raw spectra), sets of spectra were characterised qualitatively by observing the positive and negative peaks, which occur at specific wavelengths [23]. Positive peaks are due to the component of interest, while negative peaks correspond to interfering components [31]. Due to the presence of the same spectrally active properties in all locations, the Vis-NIR spectra of all soil sample sets were similar. The characteristic wavebands of reflectance spectra were only around 1400, 1900, and 2200 nm. However, there were more features of high variability at around 460–550, 1400, 1900–2000 and 2200 nm in the first derivative. The second derivative showed a similar spectrum in all locations. Stenberg et al. [32] indicated that the first and second derivatives were by far the most popular spectral preprocessing techniques for soil property prediction using Vis and NIR spectroscopy.

From Fig. 2, it is clear that three essential absorption bands are evident throughout all the compressed spectra (around 1400, 1900 and 2200 nm). Also, the general shape and slopes of all the curves are similar. The regions around 1400 and 1900 nm were related to vibrational frequencies of OH groups in the water and hydroxyl absorption, and the features around 2000–2500 nm were related to the characteristics of soil organic matter and clay minerals [6, 33]. According to Song et al. [10], although intense bands in the Vis-NIR spectra are not directly associated to the presence of PTEs or other constituents of interest in this paper, it is clear that PTEs can interact with the main spectrally active components of soil. Based on this phenomenon, chemometric models can be developed for soil samples in order to screen their toxic element concentrations. Similar results were reported by Ben-Dor et al. [34], Janik et al. [35], Kooistra et al. [36, 37], Wu et al. [3], Ren et al. [6] and Song et al. [10].

Matrix Correlation of PTEs and Reflectance Spectra

An easy approach to visualize spectral implications of PTEs is to plot correlation spectra (i.e. the correlation between the attributes and measured reflectance for each wavelength). As was shown by Song et al. [10], linear correlation coefficients between reflectance and PTEs were low to moderate. For example, in Březno was observed-0.6 < r <

0.6 throughout the Vis and NIR regions (Fig. 3). However, this indicates that PTEs contribute to the reflectance of almost all wavelengths. Fig. 3 also shows that the concentrations of toxic elements in six dumpsite soil samples displayed complex changes in their correlations with the Vis-NIR reflectance of soil spectra. Moreover, it can be seen that each element exhibits its maximum correlation coefficient at a different wavelength. Correlation analysis also indicates that the correlation coefficients of Cd and Pb, are usually separated from the other elements. Wu et al. [38] obtained the same results; they related this to week correlation of these elements (Cd, Pb) with spectra and Fe.

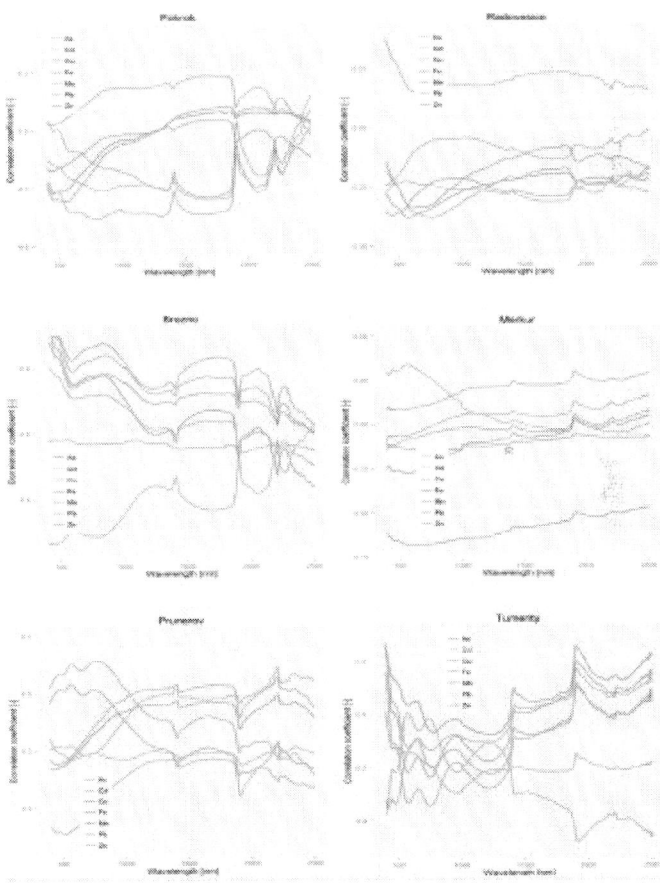

Figure 3: Correlation between reflectance of Vis-NIR and PTEs in different locations.

At each location, the PTEs were categorized into two or three groups according to their behavior and relationships with soil Vis-NIR spectra. This grouping is useful to more easily recognize the prediction ability of Vis-NIR spectroscopy for PTEs with similar behavior. Moreover, it simplifies the estimation of prediction accuracy of each PTE in a group. In Pokrok, PTEs were categorized into two groups. The first group of elements (Cd, Pb and As) had stronger negative correlation coefficients with spectral bands than the second group (Cu, Zn, Fe and Mn); the first group displayed relatively moderate negative spectral correlation at 786 nm for Pb, and the second group had the highest correlation at 1667 nm for Zn (Fig. 3). However, correlation coefficient changes of PTEs with Vis-NIR spectra of the Prunéřov dumpsite, in contrast to dumpsite Pokrok categorized into three groups, namely (Mn, As and Fe), (Zn and Cu) and (Cd and Pb), but it also displayed the strongest correlations for Pb (at 513 nm) and for Zn (at 769 nm).

In Radovesice, the highest positive, and also the highest negative, spectral correlations can both be seen in the first group of elements (Zn, Cu, As and Fe), in which the strongest positive correlation coefficient related to Zn at 401 nm. Fe represented the lowest spectral correlation at 578 nm arising from Fe^{3+} absorption. These results were similar to results of Ben-Dor [39]. They mentioned that the contribution of the region 390–550 nm is attributed to the spectral absorption features of free iron oxides. The Březno dumpsite also exhibited the highest positive and negative correlation coefficients in the Vis region, at 401 nm (Zn) and 433 nm (Cd), respectively. Correlation coefficient changes of all PTEs at the Merkur dumpsite exhibited similar behavior to Vis-NIR spectra, and categorized into one group. Similar to results of Vohland et al. [40], the order of correlation coefficients between the PTEs in this dumpsite was Cd > Zn > As > Pb > Mn > Cu > Fe, and the highest positive and negative correlation coefficients again belonged to the Vis region (561 nm to 651 nm, respectively).

Clearly, correlation changes in the Vis region of the first group of PTEs (Fe, As, Cu and Zn) in the Tumerity dumpsite fell into the 410–540 nm range. Correlation changes of Cd, which categorized into the second group of PTEs (Cd and Pb); fell into the NIR region with the highest negative correlation coefficient at 1913 nm (around the water absorption band). Stenberg et al. [32] also reported the same results.

An earlier report by Song et al. [10] indicated similar results for agricultural soils of the Changjiang River Delta, China. These findings provide support for the use of diffuse reflectance spectra in predicting the PTEs contents of soil samples.

Active soil constituents such as SOM and soil texture, as well as the indirect relationship between chromophores and PTEs can affect the reflectance of PTEs and their correlation with whole reflectance spectra [5, 10, 34]. For example, Kooistra et al. [36] found that there was a positive correlation between the SOM content and the contents of Zn and Cd in floodplains along the river Rhine in the Netherlands, and that increasing SOM content, reflectance of Zn and Cd was changed. Song et al. [10] also mentioned that the wavelength bands with highest correlation for Pb and Cd should correspond to SOM, which suggested that associations with SOM may be the main form of Pb and Cd binding in soils. Their results also showed that Cr, Cu and As had stronger negative correlation coefficients with the spectral bands attributable to the absorption features of clay and organic matter, suggesting that they are strongly bound to these soil constitutes [10].

Multivariate Analysis Using PLSR and SVMR and Validation Test

The first derivative technique was selected as the most suitable preprocessing technique [21]. Multivariate calibration techniques such as PLSR and SVMR have been used to extract soil PTEs calibration models from the reflectance spectra of soils in the Vis and NIR. The adequacy of each calibration model was evaluated based on the value of R^2_{cv}, $RMSEP_{cv}$ and the RPD [41].

As can be seen in Table 2, the two modeling strategies considered in this study provide different prediction accuracies of the studied PTEs. For the PLSR calibration set, R^2_{cv} values ranged between 0.44 and 0.61. Good and excellent R^2_{cv} ($R^2_{cv} > 0.81$ and $R^2_{cv} > 0.90$, respectively) [28] were not obtained for any of the studied elements. The best predictive models were obtained for As ($R^2_{cv} = 0.61$, $RMSEP_{cv} = 2.98$, RPD = 1.81), followed by Cd ($R^2_{cv} = 0.57$, $RMSEP_{cv} = 0.11$, RPD = 1.68). Inadequate models (high $RMSEP_{cv}$ and low R^2_{cv} and RPD) were obtained for Mn, Zn and Fe. The large variability of the sample set (colour and texture of samples from different dumpsites) used in this study also affects the accuracy of PLSR calibration models developed for PTEs.

Table 2: Statistical results for calibration and cross-validation of Vis-NIR diffuse reflectance spectroscopy for each PTEs

PTE	n		PLSR			SVMR	
		R2,CV	RMSEPCV	RPD	R2CV	RMSEPCV	RPD
Cu	115	0.50	628	1.45	0.78	4.08	229
Mn	264	0.44	116.43	1.45	0.58	101.25	1.75
Fe	264	0.48	1619.03	1.32	0.71	1141.08	2.04
Cd	104	0.57	0.11	1.68	0.78	0.08	231
Pb	148	0.51	3.12	1.50	0.66	224	1.97
Zn	115	0.45	21.84	1.42	0.71	14.51	2.16
As	104	0.61	2.98	1.81	0.89	1.89	2.63

doi:10.1371/journal.pone.0117457.t002

Kooistra et al. [37] predicted Cd and Zn contents in the floodplains of the river Rhine in the Netherlands using high-resolution reflectance spectra based on laboratory measurements with 69 soil samples. They reported very good and satisfactory predictions for both Cd (R^2_{cv} = 0.94, $RMSEP_{cv}$ = 0.68) and Zn (R^2_{cv} = 0.95, $RMSEP_{cv}$ = 80.974). Wu et al. [3] reported approximate and acceptable predictions for Cu (R^2_{cv} = 0.79, $RMSEP_{cv}$ = 6.01), Pb (R^2_{cv} = 0.81, $RMSEP_{cv}$ = 5.3), Zn (R^2_{cv} = 0.79, $RMSEP_{cv}$ = 12.83) and also As (R^2_{cv} = 0.65, $RMSEP_{cv}$ = 1.23) in soils of the Nanjing area, China. In the study of Xie et al. [5], the models provided fairly accurate predictions for Fe (R^2_{cv} > 0.80, $RMSEP_{cv}$ = 2.74, RPD > 2.00), less accurate, but acceptable for screening purposes for Cu, Pb, and Cd (0.50 < R^2_{cv} < 0.80, $RMSEP_{cv}$ = 44.05, 4.66 and 0.41 respectively, 1.40 < RPD < 2.00) and poor accuracy for Zn (R^2_{cv} < 0.50, $RMSEP_{cv}$ = 5.84, RPD < 1.40). Due to this variability, researchers tended to develop calibration models for each field that they measured with Vis and NIR spectroscopy [42, 43]. Moreover, Dunn et al. [44] indicated that the poor predictive ability of Vis and NIR for many soil constituents might result from a poorly distributed sample set with a small range, rather than the inability of Vis and NIR to predict the soil property. Besides sample variation, sample distribution and sample size are all critical to a successful Vis-NIR calibration [5].

To the best of our knowledge, the SVMR technique has not yet been commonly used to analyse and predict PTEs in the spectra domain. In the current work, SVMR was also used to develop prediction models. The results of the SVMR model for Cu, Mn, Fe, Cd, Pb, Zn and As in Vis-NIR spectra are shown in Table 2. Among the studied PTEs, As is the most accurately predicted with SVMR (R^2_{cv} = 0.89, $RMSEP_{cv}$ = 1.89, RPD = 2.63). This prediction accuracy is classified to be very good. The calibration results for Mn were not as good as the results of the other elements (R^2_{cv} = 0.58, $RMSEP_{cv}$ = 101.25, RPD = 1.75), indicating a fair model. Results obtained with SVMR for Cu, Fe, Cd, Pb and Zn gives a good classification, although the prediction accuracy of Fe and Pb is slightly lower than those of Cu and Cd.

Table 2 indicates that in the validation procedure, cross-validation R^2 (R^2_{cv}) of PLSR ranged between 0.44 for Mn and 0.61 for As, while the range for SVMR was between 0.58 for Mn to 0.89 again for As. Based on R^2_{cv} and $RMESP_{cv}$, which has been introduced as standard methods for validation of the prediction models, in both calibration and validation, the best estimates were clearly obtained for As prediction. Generally, R^2_{cv} and $RMESP_{cv}$ and also RPD for both methods were satisfactory, but the same as for calibration, SVMR results were more reliable which emphasizes the need for using more flexible techniques, such as SVMR.

By comparing the results of the PLSR and SVMR models for the Vis-NIR spectra, it can be seen that the use of PLSR has been generally successful to calibrate many soil variables including some PTEs concentrations [5, 10, 45]. In this study, PLSR showed moderate predictions, however; SVMR provided fairly good correlations between soil spectra and various PTEs; better prediction was achieved using SVMR, which outperformed the PLSR. From a practical point of view, the prediction accuracies obtained with these two methods generally seem to be acceptable for a number of agricultural applications, including soil science research. The superior performance of SVMR over PLSR can be explained by the inclusion of nonlinear and interaction effects, as well as linear combinations of variables. It is able to approximate nonlinear functions between multidimensional spaces [7, 46, 47].

CONCLUSIONS

This study demonstrated the application of laboratory Vis-NIR reflectance spectroscopy for the prediction of PTEs, including Cu, Mn, Cd, Zn, Fe, Pb and As, using soil samples taken from six brown coal mining dumpsites of the Czech Republic. For each parameter, Vis-NIR calibration models were created by PLSR and SVMR algorithms. Correlation analysis revealed that PTEs contribute to the reflectance of almost all wavelengths, and that the patterns of correlation coefficients of Cd and Pb are usually separated from the other elements. The results showed obvious differences in predictability and accuracy of PLSR and SVMR. When using a SVMR model, soil spectroscopy was shown to be a very promising method for the determination of PTE concentrations in anthropogenic soils. The best predictability of Vis-NIR reflectance spectroscopy was obtained by SVMR for As ($R^2 > 0.90$, $RMSEP_{cv} =$ 1.89, RPD > 2.5), followed by Cd, Cu, Zn, Pb and Mn. In generally, our results confirmed that Vis-NIR reflectance spectroscopy combined with first derivative and SVMR methods have great potential for site-specific soil monitoring in high-risk regions. It leads to overoptimistic performance in the assessment of PTEs, which generally involves conducting large numbers of analyses in a short time. For future investigations, hyperspectral sensors may be useful, and have to be explored for fitting specific spectral regions and for models to optimize the estimation of PTEs content.

ACKNOWLEDGMENTS

The authors acknowledge the assistance of Mr. Christopher Ash for English revision.

AUTHOR CONTRIBUTIONS

Conceived and designed the experiments: AG MMS. Performed the experiments: AK JK Analyzed the data: AG RV. Contributed reagents/materials/analysis tools: AG LB OD VT Wrote the paper: AG MMS LB

REFERENCES

1.   Li MS (2006) Ecological restoration of mineland with particular reference to the metalliferous mine wasteland in China: A review of research and practice. Sci Total Environ 357:38–53. pmid:15992864 doi: 10.1016/j.scitotenv.2005.05.003

2.   Strudl M, Boruvka L, Dimitrovsky K, Kozak J (2006) Content of potentially risk elements in natural and reclaimed soils of Sokolov region. Soil Water Res. 1:99–107.

3.   Wu Y, Chen J, Wu X, Tian Q, Ji J, et al. (2005) Possibilities of reflectance spectroscopy for the assessment of contaminant elements in suburban soils. Appl Geochem 20:1051–1059. doi: 10.1016/j.apgeochem.2005.01.009

4.   N'Guessan YM, Probst JL, Bur T, Probst A, (2009) Trace elements in stream bed sediments from agricultural catchments (Gascogne region, S-W France): where do they come from? Sci Total Environ 407:2939–2952. doi: 10.1016/j.scitotenv.2008.12.047. pmid:19215965

5.   Xie X, Pan XZ, Sun B (2012) Visible and near-infrared diffuse reflectance spectroscopy for prediction of soil properties near a Copper smelter. Pedosphere 22:351–366. doi: 10.1016/s1002-0160(12)60022-8

6.   Ren HY, Zhuang DF, Singh AN, Pan JJ, Qid DS, et al. (2009) Estimation of As and Cu contamination in agricultural soils around a mining area by reflectance spectroscopy: A case study. Pedosphere 19:719–726. doi: 10.1016/s1002-0160(09)60167-3

7.   Zornoza R, Guerrero C, Mataix-Solera J, Scow KM, Arcenegui V, et al. (2008) Near infrared spectroscopy for determination of various physical, chemical and biochemical properties in Mediterranean soils. Soil Biol Biochem 40:1923–1930. pmid:23226882 doi: 10.1016/j.soilbio.2008.04.003

8.   Gholizadeh A, Boruvka L, Saberioon MM, Vasat R (2013) Visible, near-infrared, and mid-infrared spectroscopy applications for soil assessment with emphasis on soil organic matter content and quality: State-of-the-art and key issues. Appl Spectrosc 67:1349–1362. doi: 10.1366/13-07288. pmid:24359647

9.   Salazara DM, Martinez Reyesa HL, Martinez-Rosasa ME, Miranda Velascoa MM, Arroyo Ortegaa E, et al. (2012) Visible-near infrared spectroscopy to assess soil contaminated with cobalt. Procedia Eng 35:245–253.

10.   Song Y, Li F, Yang Z, Ayoko GA, Frost RL, et al. (2012) Diffuse reflectance spectroscopy for monitoring potentially toxic elements in the agricultural soils of Changjiang River Delta, China Appl Clay Sci 64:75–83. doi: 10.1016/j.clay.2011.09.010

11. Kemper T, Sommer S (2002) Estimate of heavy metal contamination in soils after a mining accident using reflectance spectroscopy. Environ Sci Technol 36:2742–2747. pmid:12099473 doi: 10.1021/es015747j

12. Bray JGP, Viscarra Rossel RA, McBratney AB (2009) Diagnostic screening of urban soil contaminants using diffuse reflectance spectroscopy. Aust J Soil Res 47:433–442. doi: 10.1071/sr08068

13. Martens H, Naes T (1989) Multivariate calibration. John Wiley and Sons, New York, p. 419. pmid:2612913

14. Dalal RC, Henry RJ (1986) Simultaneous determination of moisture, organic carbon, and total nitrogen by near infrared reflectance spectrophotometry. Soil Sci Soc Am J 50:120–123. doi: 10.2136/sssaj1986.03615995005000010023x

15. Pirie A, Singh B, Islam K (2005) Ultra-violet, visible, near-infrared, and mid infrared diffuse reflectance spectroscopic techniques to predict several soil properties. Aust J Soil Res 43:713–721. doi: 10.1071/sr04182

16. Moros J, de Vallejuelo SFO, Gredilla A, de Diego A, Madariaga JM, et al. (2009) Use of reflectance infrared spectroscopy for monitoring the metal content of the estuarine sediments of the Nerbioi-Ibaizabal River (Metropolitan Bilbao, Bay of Biscay, Basque Country). Environ Sci Technol 43:9314–9320. doi: 10.1021/es9005898. pmid:20000524

17. Vapnik V (1995) The Nature of Statistical Learning Theory. Springer-Verlag, New York.

18. McGrath SP, Cunliffe CH (1985) A simplified method for the extraction of metals Fe, Zn, Cu, Ni, Cd, Pb, Cr, Co and Mn from soils and sewage sludges. J Sci Food Agr 36:794–798. doi: 10.1002/jsfa.2740360906

19. .Gomez C, Lagacherie P, Coulouma G (2012) Regional predictions of eight common soil properties and their spatial structures from hyperspectral Vis-NIR data. Geoderma 189–190: 176–185. doi: 10.1016/j.geoderma.2012.05.023

20. Murray I (1988) Aspects of interpretation of NIR spectra, in: Creaser C.S., Davies A.M.C. (Eds.), Analytical Application of Spectroscopy. Royal Society of Chemistry, London, pp. 9–21.

21. Duckworth J (2004) Mathematical data preprocessing, in: Roberts C.A., Workman J. Jr., Reeves J.B. III (Eds.), Near-Infrared Spectroscopy in Agriculture, ASA-CSSA-SSSA, Madison, WI pp. 115–132.

22. Rinnan A, van den Berg F, Engelsen SB (2009) Review of the most common pre-processing techniques for near-infrared spectra. Trend Anal Chem 28:1201–1222. doi: 10.1016/j.trac.2009.07.007

23. Viscarra Rossel RA, Walvoort DJJ, McBratney AB, Janik LJ, Skjemstad JO (2006a) Visible, near-infrared, mid-infrared or combined diffuse reflectance spectroscopy for simultaneous assessment of various soil properties. Geoderma 131:59–75. doi: 10.1016/j.geoderma.2005.03.007

24. Mevik BH, Wehrens R (2007) The pls package: principal component and partial least squares regression in R. J Stat Softw 18:1–24.

25. Vohland M, Besold J, Hill J, Frund HC (2011) Comparing different multivariate calibration methods for the determination of soil organic carbon pools with visible to near infrared spectroscopy. Geoderma 166:198–205. doi: 10.1016/j.geoderma.2011.08.001

26. Venkoba Rao B, Gopalakrishna SJ (2009) Hardgrove grindability index prediction using support vector regression. Int J Miner Process 91:55–59. doi: 10.1016/j.minpro.2008.12.003

27. Meyer D, Dimitriadou E, Hornik K, Weingessel A, Leisch F (2012) e1071: Misc Functions of the Department of Statistics (e1071), R package version 1.6–1, Wien.

28. Williams P (2003) Near-infrared technology-Getting the best out of light, PDK Projects, Nanaimo, Canada.

29. Mouazen AM, Kuang B, De Baerdemaeker J, Ramon H (2010) Comparison between principal component, partial least squares and artificial neural network analyses for accuracy of measurement of selected soil properties with visible and near infrared spectroscopy. Geoderma 158:23–31. doi: 10.1016/j.geoderma.2010.03.001

30. Viscarra Rossel RV, McGlyn RN, McBratney AB (2006b) Determining the composition of mineral-organic mixes using UV-Vis-NIR diffuse reflectance spectroscopy. Geoderma 137:70–82. doi: 10.1016/j.geoderma.2006.07.004

31. Haaland DM, Thomas EV (1988) Partial least-squares methods for spectral analyses: 1. Relation to other quantitative calibration methods and the extraction of qualitative information. Anal Chem 60:1193–1202. doi: 10.1021/ac00162a020

32. Stenberg B, Viscarra Rossel RA, Mouazen AM, Wetterlind J (2010) Visible and near infrared spectroscopy in soil science, in: Donald L. (Eds.), Advances in Agronomy, Burlington, VT: Elsevier, pp. 163–215.

33. Kooistra L, Wanders J, Epema GF, Leuven R, Wehrens R, et al. (2003) The potential of field spectroscopy for the assessment of sediment properties in river floodplains. Anal Chim Acta 484:189–200. doi: 10.1016/s0003-2670(03)00331-3

34. Ben-Dor E, Inbar Y, Chen Y (1997) The reflectance spectra of organic matter in the visible near-infrared and short wave infrared region (400–2500 nm) during a controlled decomposition process. Remote Sens Environ 61:1–15. doi: 10.1016/s0034-4257(96)00120-4

35. Janik LJ, Merry RH, Skjemstad JO (1998) Can mid infra-red diffuse reflectance analysis replace soil extractions? Aust J Exp Agr 38:681–696. doi: 10.1071/ea97144

36. Kooistra L, Wehren R, Buydens LMC, Leuven RSE, Nienhuis PH (2001a) Possibilities of soil spectroscopy for the classification of contaminated areas in river floodplains. Int J Appl Earth Obs Geoinform 3:337–344. doi: 10.1016/s0303-2434(01)85041-8

37. Kooistra L, Wehren R, Leuven RSE, Buydens LMC (2001b) Possibilities of visible-near-infrared spectroscopy for the assessment of soil contamination in river flood plains. Anal Chim Acta 446:97–105. doi: 10.1016/s0003-2670(01)01265-x

38. Wu Y, Chen J, Ji J, Gong P, Liao Q, et al. (2007) A mechanism study of reflectance spectroscopy for investigating heavy metals in soils. Soil Sci Soc Am J 71:918–926. doi: 10.2136/sssaj2006.0285

39. Ben-Dor E (2002) Quantitative remote sensing of soil properties. Adv Agron 75:173–243. doi: 10.1016/s0065-2113(02)75005-0

40. Vohland M, Bossung1 B, Frund HC (2009) A spectroscopic approach to assess trace-heavy metal contents in contaminated floodplain soils via spectrally active soil components. J Plant Nutr Soil Sci 172:201–209. doi: 10.1002/jpln.200700087

41. Mouazen AM, Maleki MR, De Baerdemaeker J, Ramon H (2007) On-Line measurement of some selected soil properties using a VIS—NIR sensor. Soil Till Res 93:13–27. doi: 10.1016/j.still.2006.03.009

42. Imade Anom SW, Shibusawa S, Sasao A, Sato H, Hirako S, et al. (2000) Moisture, soil organic matter and nitrate nitrogen content maps using the real-time soil spectrophotometer, in: IFAC Bio-robotics, information technology and intelligent control for bio-production systems, Sakai, Osaka, Japan, pp. 307–312.

43. Mouazen AM, De Baerdemaeker J, Ramon H (2005) Towards development of on-line soil moisture content sensor using a fibre-type NIR spectrophotometer. Soil Till Res 80:171–183. doi: 10.1016/j.still.2004.03.022

44. Dunn BW, Beecher HG, Batten GD, Ciavarella S (2002) The potential of near-infrared reflectance spectroscopy for soil analysis-a case study from the Riverine Plain of south-eastern Australia. Aust J Exp Agr 42:607–614.

45. Gannouni S, Rebai N, Abdeljaoued S (2012) A Spectroscopic approach to assess heavy metals contents of the mine waste of Jalta and Bougrine in the north of Tunisia. J Geogr Inf Syst 4:242–253. doi: 10.4236/jgis.2012.43029

46. Bilgili AV, van Es HM, Akbas F, Durak A, Hively WD (2010) Visible-near infrared reflectance spectroscopy for assessment of soil properties in a semi-arid area of Turkey. J Arid Environ 74:229–238. doi: 10.1016/j.jaridenv.2009.08.011

47. Stevens A, Udelhoven T, Denis A, Tychon B, Lioy R, et al. (2010) Measuring soil organic carbon in croplands at regional scale using airborne imaging spectroscopy. Geoderma 158:32–45. doi: 10.1016/j.geoderma.2009.11.032

Integrated Assessment of Heavy Metal Contamination in Sediments from a Coastal Industrial Basin, NE China

Xiaoyu Li[1, 2], Lijuan Liu[1], Yugang Wang[1], Geping Luo[1], Xi Chen[1], Xiaoliang Yang[3], Bin Gao[4], and Xingyuan He[2]

[1]State Key Laboratory of Desert and Oasis Ecology, Xinjiang Institute of Ecology and Geography, Chinese Academy of Sciences, Xinjiang, China

[2]State Key Laboratory of Forest and Soil Ecology, Institute of Applied Ecology, Chinese Academy of Sciences, Liaoning, China

[3]College of Environmental Science and Forestry, State University of New York, Syracuse, New York, United States of America

[4]College of Resources Science and Technology, Beijing Normal University, Beijing, China

ABSTRACT

The purpose of this study is to investigate the current status of metal pollution of the sediments from urban-stream, estuary and Jinzhou Bay of the coastal industrial city, NE China. Forty surface sediment samples from river, estuary and bay and one sediment core from Jinzhou bay were collected and analyzed for heavy metal concentrations of Cu, Zn, Pb, Cd, Ni and Mn. The data reveals that there was a remarkable change in the contents of heavy metals among the sampling sediments, and all the mean values of heavy metal concentration were higher than the national guideline values of marine sediment quality of China (GB 18668-2002). This is one of the most polluted of the world's impacted coastal systems. Both the correlation analyses and geostatistical analyses showed that Cu, Zn, Pb and Cd have a very similar spatial pattern and come from the industrial activities, and the concentration of Mn mainly caused by natural factors. The estuary is the most polluted area with extremely high potential ecological risk; however the contamination decreased with distance seaward of the river estuary. This study clearly highlights the urgent need to make great efforts to control the industrial emission and the exceptionally severe heavy metal pollution in the coastal area, and the immediate measures should be carried out to minimize the rate of contamination, and extent of future pollution problems.

INTRODUCTION

Coastal and estuarine areas are among the most important places for human inhabitants [1]; however, with rapid urbanization and industrialization, heavy metals are continuously carried to the estuarine and coastal sediments from upstream of tributaries [2]–[5]. Heavy metal contamination in sediment could affect the water quality and bioaccumulation of metals in aquatic organisms, resulting in potential long-term implication on human health and ecosystem [6]–[7]. In most circumstances, the major part of the anthropogenic metal load in the sea and seabed sediments has a terrestrial source, from mining and industrial developments along major rivers and estuaries [8]–[10]. The hot spots of heavy metal concentration are often near industrial plants [11]. Heavy metal emissions have been declining in some industrialized countries over the last few decades [12], [13], however, anthropogenic sources have been increasing with rapid industrialization and urbanization in developing countries [14], [15].Heavy metal contaminations in sediment could affect the water quality, the bio assimilation and bioaccumulation of metals in aquatic organisms, resulting in potential long-term effects on human health and ecosystem [16]–[19]. Quantification of the land-derived metal fluxes to the sea is therefore a key factor to ascertain at which extent those inputs can influence the natural biogeochemical processes of the elements in the marine [20], [21]. The spatial distribution of heavy metals in marine sediments is of major importance in determining the pollution history of aquatic systems [22], [23], and is basic information for identifying the possible sources of contamination and to delineate the areas where its concentration exceeds the threshold values and the strategies of site remediation [24]. Therefore, understanding the mechanisms of accumulation and geochemical distribution of heavy metals in sediments is crucial for the management of coastal environment.

China's rapid growth of the economy since 1979 under the reform policies has been accompanied by considerable environmental side effects [25]. China is one of the largest coastal countries in the world. Booming coastal urban areas are increasingly dumping huge industrial and domestic waste at sea [26]. The elevated metal discharges put strong pressure on China's costal and estuarine area. The average annual input of metals by major rivers was approximately 30,000 t between 2002 and 2008 [27]. Chinese government indicates that 29,720 km² of offshore areas of China are heavily polluted [28]. "Hot spots" of metal contamination can be found along the coast of China [29], from the north to the south, especially in the industry-developed estuaries, such as the Liaodong Bay [30] and Yangtze River catchment [31] and Xiamen Bay [32]. In 2002, China enforced Marine Sediment Quality (GB 18668-2002) to protect marine environment (CSBTS, 2002). Therefore, Marine Sediment Quality (GB 18668-2002) is used as a general measure of marine sediment contamination in China.

Jinzhou Bay, surrounded by highly industrialized regions, is considered as one of the most contaminated coastal areas in China [33]. China produces the largest amount of zinc (Zn) in the world, which was 1.95 million tons in 2000 and will grow to 14.9 million tons in 2010 [26], [34]. And the largest zinc smelting plant in Asia was located at the coast of Jinzhou bay. From 1951 to 1980, the amount of Zn, Cu, Pb and Cd discharged from Huludao Zinc Smelter to Jinzhou bay reached 33745, 3689, 3525 and 1433 t respectively [35]. Although several heavy metal contamination studies have conducted in Jinzhou Bay area recently, these studies were focused on coastal urban soils [36], river sediments [37] and seawater [38] separately. Few researches take the coastal stream, estuary and bay as a whole unit to assess the heavy metal contamination of coastal industrial area spatially and temporally. Thus, it is necessary to understand the process of heavy metal contamination and to evaluate the potential ecological risks of heavy metals in the coastal stream, estuary and bay integratively.

In recent decades different metal assessment indices applied to sediment environments have been developed. Caeiro et al [9] classified them in three types: contamination indices, background enrichment indices and ecological risk indices. The geo-accumulation index (I_{geo}) [39] and the potential ecological risk index (RI) [40] are the most popular methods used to evaluate the ecological risk posed by heavy metals in sediments [41]–[44]. RI method considers the toxic-response of a given substance and the total risk index, and can exhibit the actual pollution condition of seriously polluted sediment [45], [46].

Over the last few decades the study of the sediment cores has shown to be an excellent tool for establishing the effects of anthropogenic and natural processes on depositional environments[44], [47]. Sediment cores can be used to study the pollution history of aquatic ecosystem [48],[49]. Within an individual sediment core, differences in pollutant concentrations at different depths reflect how heavy metal input and accumulation changes over time [50], [51].

The purpose of this study is (1) to quantify and explain the spatial distribution of heavy metal contaminants in modern sediments of Jinzhou bay, NE China; (2) to investigate the natural and anthropogenic processes controlling sediment chemistry; and (3) to identify the potential ecological risks of such heavy metals.

MATERIALS AND METHODS

Study Area

This study was carried out in Jinzhou Bay and its coastal city, Huludao City, in Liaoning Province, northeast of China (Fig. 1). Jinzhou Bay is one of the important bays in the northwest of Liaodong Bay at the northwestern bank of China's Bohai Sea. It is a semi-closed shallow area with an average depth of 3.5 m and an approximate area of 120 km^2. Huludao city is located at southwestern coast of Jinzhou bay. The city is an important non-ferrous smelting and chemical industry area in northeast China. More than forty different mineral resources have been discovered in the Huludao region, including gold, zinc, molybdenum, lime and manganese. The economy is dominated by some of China's most important industrial enterprises, such as Asia's biggest zinc manufacturing operation, the Huludao Zinc Smelter (HZS), the Jinxi Oil Refinery and Jinhua Chemical Engineering, and Huludao's Massive Shipyard. The Wuli River, Lianshan River and the Cishan River are three main rivers in the city, flowing into Jinzhou Bay. The water, soil and sediment in the city and Jinzhou bay were heavily polluted by industrial activities. Land reclamation from sea by landfill of soils and solid wastes further increase the level of pollution of the sedimentary environments in this area. These anthropogenic activities have created great threat to the public health and the regional biological and geochemical conditions.

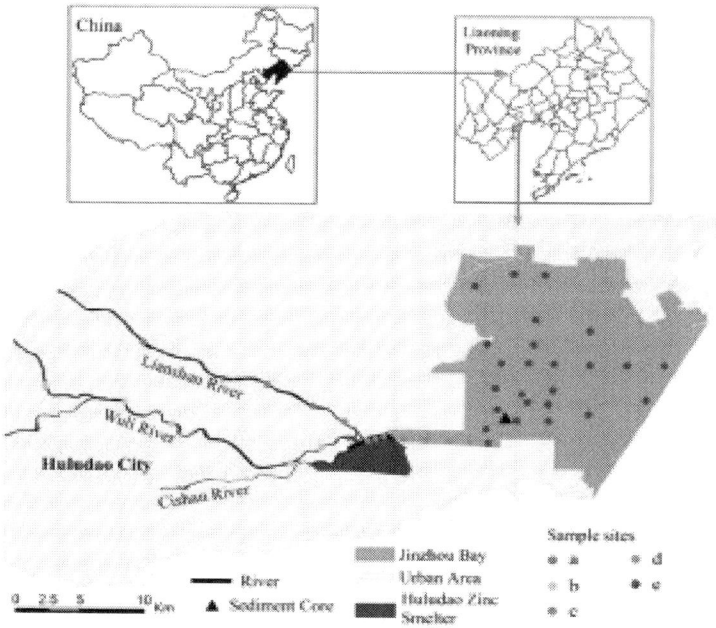

Figure 1: The sampling sites in the study area. (a, Lianshan River upstream of Huludao Zinc Smelter (n = 4); b, Wuli River upstream of Huludao Zinc Smelter (n = 5); c, Converged river Downstream of Huludao Zinc Smelter (n = 3); d, Estuary (n = 4); e, Jinzhou Bay (n = 25)).

No specific permits were required for the described field studies. The studying area is not privately-owned or protected in any way and the field studies did not involve endangered or protected species.

Sampling and Analysis

Twelve samples of river sediment were collected from the two major rivers (Lianshan river and Wuli river) and and four samples were collected from their estuary of Huludao City, using a stainless steel shovel. Twenty-five of surface sediments (0–5 cm) and one sediment core were collected in Jinzhou Bay using a stainless gravity corer (40 cm length and 5 cm diameter). The sediment core was sectioned at 2 cm intervals, and each fraction (subsamples) was sliced into 50 ml polyethylene centrifuge tubes with the help of PVC spatula. All the samples were collected in October 2009 in one week.

The samples were oven-dried at 45°C for 3 days, and sieved through a 2-mm plastic sieve to remove large debris, gravel-size materials, plant roots and other waste materials, and stored in closed plastic bags until analysis. Soil was digested with a mixture 5:2:3 of HNO_3–$HClO_4$–HF. The digested solutions were analyzed via an inductively coupled plasma-atomic emission spectroscopy (ICP-AES; Perkin Elmer Optima 3300 DV). All of the soil samples were analyzed for total concentrations of Cu, Zn, Pb, Ni, Mn and Cd.

Statistical Analysis

Statistical methods were applied to process the analytical data in terms of its distribution and correlation among the studied parameters. The commercial statistics software package SPSS version 17.0 for Windows was used for statistical analyses in present study. Basic statistical parameters such as mean, median, standard deviation (SD), coefficient of variation (CV), skewness and kurtosis were computed. To identify the relationship among heavy metals in sediments and their possible sources, Pearson's correlation coefficient analysis were performed.

Geostatistical Methods

Semivariogram is a basic tool of geostatistics and also the mathematical expectation of the square of regional variable $z(x_i)$ and $z(x+h_j)$ increment, namely the variance of regional variable. Its general form is:

$$\gamma(h) = \frac{1}{2N(h)} \sum_{i=1}^{N(h)} [z(x_i) - z(x_i + h)]^2$$

where $r(h)$ is semivariogram; h is step length, namely the spatial interval of sampling points used for the classification to decrease the individual number of spatial distance of various sampling point assemblages; $N(h)$ is the logarithm of sampling point when the spacing is h; $z(x_i)$ and $z(x_i+h)$ are the values when the variable Z is at the x_i and x_i+h positions respectively. The residual sums of squares (RSS), the determining coefficient (R^2) and F test were used to evaluate the accuracy of the interpolated results.

Kriging, as a geostatistical interpolation method, uses the semivariogram to quantify the spatial variability of regionalized variables, and provides parameters for spatial interpolation. The maps of spatial distribution of heavy metal concentrations were generated by Kriging interpolation with the support of the statistical module of ArcGIS-Geostatistical Analyst.

Potential Ecological Risk

To assess the effect of multiple metal pollutions in the sediments from the river, estuary and Jinzhou bay, potential ecological Risk Index (RI) was used, which was originally developed by Håkanson [40] and is widely used in ecological risk assessments of heavy metals in sediments. According to this methodology, the potential ecological risk index (RI) is defined as

$$RI = \sum Er^i \tag{1}$$

$$Er^i = Tr^i \cdot C_f^i \tag{2}$$

$$C_f^i = C_o^i / C_n^i \tag{3}$$

where RI is calculated as the sum of all risk factors for heavy metals in sediments; Er^i is the monomial potential ecological risk factor; TR^i is the toxic-response factor for a given substance (e.g., Cu = Pb = Ni = 5, Zn = 1, Cd = 30); C_f^i, C_o^i and C_n^i are the contamination factor, the concentration of metals in the sediment and the background reference level, respectively. The background values of Cu, Zn, Pb and Cd are defined as the maximum values of the first category standard of national guideline values of marine sediment quality of China (GB18668-2002) and Ni is defined as the average value of Ni in residual fraction determined, they were 35 mg/kg for Cu, 150 mg/kg for Zn, 60 mg/kg for Pb, 0.5 mg/kg for Cd, and 9 mg/kg for Ni. The concentration of Mn in the sediments showed very weak relationship with the industrial activities, so it was not included in the calculation process of RI.

Still according to Hakanson [40] the following terminology is indicated to be used for the RI value:

RI <150, low ecological risk for the sediment;

150≤ RI <300, moderate ecological risk for the sediment;

300≤ Ri <600, considerable ecological risk for sediment;

RI ≥600, very high ecological risk for the sediment

RESULTS AND DISCUSSION

Heavy Metal in the Sediments

Descriptive statistics of heavy metal concentrations of sediments present in rivers of Huludao city, estuary and Jinzhou bay (Fig. 1) are presented in Table 1, 2, 3. As confirmed by the skewness values (Table 1, 3), the concentrations of elements (except Mn) are characterized by large variability, with positively skewed frequency distributions. This is common for heavy metals, because they usually have low concentrations in the environment, so that the presence of a point source of contamination may cause a sharp increase in local concentration, exceeding the thresholds [24].

Table 1: Heavy metal concentrations (mg/kg) of River sediments

		Minimum	Maximum	Mean	Median	SD	CV%	Skewness	Kurtosis	national guideline values
Cu	a	67.00	186.50	116.50	106.25	50.22	43.11	1.15	2/4	
	b	32.15	85.50	50.61	42.40	24.55	48.51	1.45	1.73	35.00
		795.00	2535.00	1533.33	1270.00	899.39	58.66	1.20	3.51	
Zn	a	471.61	965.31	633.85	549.24	224.08	35.35	1.83	3.50	150.00
	b	153.52	473.93	256.78	199.85	146.71	57.13	1.84		
		1825.96	11010.02	6546.57	6803.73	4597.42	70.23	-.251		
Pb	a	62.64	185.10	112.28	100.69	52.34	46.62	1.18	1.76	
	b	40.15	98.47	57.60	45.89	27.47	47.69	1.90	3.66	60.00
		417.58	6090.90	2431.09	784.81	3174.79	130.59	1.70		
Ni	a	35.69	109.99	57.41	41.98	35.23	61.36	1.94	3.79	
	b	28.28	35.94	31.49	30.885	3.31	10.51	0.92	0.52	
		40.77	87.93	62.58	59.06	23.77	37.98	0.65		
Mn	a	520.89	1283.46	812.91	723.65	327.98	40.34	1.47	2.75	
	b	520.21	903.43	710.96	710.11	205.78	28.94	0.004	-5.81	
		337.93	1358.55	915.18	1049.07	523.31	57.18	-1.08		
Cd	a	25.53	98.78	53.18	44.21	32.61	61.32	1.29	1.30	
	b	8.04	17.75	11.12	9.35	4.49	40.38	1.81	3/9	0.50
		136.73	1019.10	503.51	354.71	459.62	91.28	1.30		

a Lianshan River upstream of Huludao Zinc Smelter (n=4);

[b]M/uli River upstream of Huludao Zinc Smelter (n=5);

[c]Converged river Downstream of Huludao Zinc Smelter (n = 3). The locations of sampling sites are on Fig. 1. doi:10.1371/jo uma l.pone.0039690.t001

Table 2: Heavy metal concentrations (mg/kg) of Estuary sediments

Distance from HZS to the sampling sites	Cu	Zn	Pb	Ni	Mn	Cd
400 m	1510.00	9304.24	1414.14	59.95	744.13	269.34
1200 m	805.00	3898.09	561.21	64.10	677.46	181.90
2000 m	675.00	2750.57	486.43	48.14	514.07	118.08
2800 m	505.00	3221.59	421.28	53.85	682.69	98.29

The locations of sampling sites are on Fig. 1. doi:10.1371/journal. pone.0039690.t002

Table 3: Heavy metal concentrations (mg/kg) of Jinzhou Bay sediments (n = 25)

	Minimum	Maximum	Mean	Median	SD	CV%	Skewness	Kurtosis	national guideline values
Cu	24.45	327.50	74.11	51.50	68.54	92.48	2.706	7.964	35.00
Zn	168.06	2506.33	689.39	550.58	568.50	82.46	2.183	4.656	150.00
Pb	29.17	523.45	123.98	89.29	114.70	92.51	2.398	5.934	60.00
Ni	26.29	85.99	43.47	41.41	11.92	27.42	1.835	8.867	
Mn	445.57	1123.03	750.64	774.00	153.82	20.49	0.311	0.137	
Cd	7.91	105.31	26.81	20.74	22.23	82.92	2.388	6.155	0.50

The locations of sampling sites are on Fig. 1. doi:10.1371/journal.pone.0039690.t003

Heavy Metal in River Sediments

Among the concentrations of heavy metal of river sediments (Table 1), the low values are from the sediments sampled before the river flowing by the Huludao Zinc Smelter (HZS), and the highest heavy metal concentrations come from the two sediment samples collected after the river flows by the HZS and accepted the wastewater discharged from HZS. Although the Wuli river is the least contaminated river in Huludao City; however the concentration of Cd exceeded the national guideline values of marine sediment quality of China (GB 18668-2002) by 22 times. In the upstream of HZS the sediment contamination of Lianshan river was higher than that of Wuli river. These data showed many other industrial operations at upper reaches of HZS also contributed great heavy metals to the river sediments. According to historical data, wastes from Jinxi Chemical Factory and Jinxi Petroleum Chemical Factory were discharged into Wuli River directly for nearly 40 years till 2000 and caused heavy contamination of river sediments. At the same time, wastewater from other small industrial plants and residents, and nonpoint pollution from soils with runoff or atmospheric deposition contributed additional pollution sources [52],[53]. This is be confirmed by the previous studies about contamination of urban soils [36], [54] and river sediments [37] in Huludao city.

The mean concentrations of Cu, Zn, Pb and Cd in the sediment of converged river downstream of HZS adjacent to the estuary exceeding the national guideline values of marine sediment quality of China (GB 18668-2002) by 42, 42, 40 and 1006 times respectively. These indicated that the Huludao Zinc Smelter (HZS) is the largest source of heavy metals in river sediment adjacent to the estuary. The annual amount of discharged wastewater from HZS is estimated to be more than 8 million tons directly to the Wuli river [35]. This situation lasted for more than 50 years till the water reuse was realized around year 2000 [28].

Heavy Metal in the Estuary Sediments

Four sediment samples were collected along the estuary, and the distance from HZS to the sampling sites increases from 400 m with an 800 m-interval to the Jinzhou Bay (Fig. 1). All the investigated heavy metals clearly showed the same distribution trend. The maximum concentrations of heavy metals declined as the distance increased from HZS (Table 2). The mean concentration of Cu, Zn, Pb and Cd at estuary were exceeded the national guideline values of marine sediment quality of China (GB 18668-2002) by 24, 31, 11 and 332 times respectively.

Contaminant distributions in the Hudson River estuary were identified two types of trends: increasing trend down-estuary dominated by down-estuary sources such as wastewater effluent, and decreasing trend toward bay dominated by upriver sources, where they are removed and diluted downstream along with the sediment transport [55]. The result of this study showed obvious decreasing trend toward bay. This confirmed that the wastewater discharged from HZS and other industrial plants were the main sources of heavy metals in sediment.

Heavy Metal in Jinzhou Bay Sediments

Twenty-five of surface sediments from Jinzhou Bay were tested and the minimum, maximun and mean values are all located in Table 3. There were remarkable changes in the concentrations of heavy metals in the sediments of Jinzhou bay (Fig. 1). The heavy metal concentration levels are comparable to those with previous studies [56]–[58], especially for Cu, Pb and Zn, Mean concentrations of Cu, Zn, Pb and Cd were as high as 2.1, 4.6, 2.1 and 53.6 times of the national guideline values of marine sediment quality of China (GB 18668-2002). According to the research results from Institute of Marine Environmental protection, State oceanic Administration of China in 1984, the background value of Cu, Zn, Pb and Cd in Jinzhou bay was 10.0, 48.59, 9.0 and 0.29 mg/kg respectively [35]. This clearly demonstrates an anthropogenic contribution and reveals a serious pollution of sediments in Jinzhou bay. For all metals, total concentrations had a great degree of variability, shown by the large coefficients of variation (CV) from 20.49% of Mn to 92.51% of Pb. The elevated coefficients of variations reflected the inhomogeneous distribution of concentrations of discharged heavy metals. Large standard deviations were found in all heavy metals levels. The results of the K–S test (P<0.05) showed that the concentrations of measured metals were all normally distributed.

The comparison of contaminant concentrations observed in this study with those reported for other regions (Table 4) indicates that levels and ranges of variation of our data are similar to those reported from sites with high anthropogenic impact. It was found that the concentrations of Cd measured in this study were greatly higher than other studies except that of Algeciras Bay was a little close to this study. The levels of Zn and Pd in this study were only lower than that of Izmit Bay and Gulf of Naples respectively. The contents of Cu were relatively higher than other study (except Izmit Bay, Bay of Bengal and Hong Kong). All these show that Jinzhou bay was a highly polluted area in the world.

Table 4: Mean concentrations (mg/kg) of heavy metals found in Jinzhou bay compared to the reported average concentrations for other world impacted coastal systems

Area	Cu	Zn	Pb	Ni	Mn	Cd	Reference
Izmit Bay, Turkey	89.4	754	94.9	52.1	-	6.3	[56]
Ribeira Bay, Brazil	24.6	109	22.9	47	466	0.207	[59]
Sepetiba Bay, Brazil	31.9	567	40	22.3	595	3.22	[59] Mil
Mejillones Bay,Chile		29.7	-	20.6	93.8	21.9	[58]
Algeciras Bay, Spain	17	73	24	65	534	0.3	[60]
Taranto Gulf, Italy	47.4	102.3	57.8	53.3	893		[11]
Tivoli South Bay,USA	17.6	92.8	26.3	-		-	[61]
Bay of Bengal, India	677.7	60.39	25.66	34.03	366.66	5.24	[57]
Gulf of Mannar, India	57	73	16	24	305	0.16	[62]

Gulf of Naples, Italy	27.2	602	221	6.93	1550	0.57	[63]
Hong Kong, China	118.68	147.73	53.56	24.72	523.99	0.33	[64]
This study	74.11	689.39	123.98	43.47	750.64	26.81	

Note: "–" = no data. doi:10.1371/journal.pone.0039690.t004

Correlation between Heavy Metals

Correlation analyses have been widely applied in environmental studies. They provided an effective way to reveal the relationships between multiple variables in order to understand the factors as well as sources of chemical components [50], [65]. Heavy metals in environment usually have complicated relationships among them. The high correlations between heavy metals may reflect that the accumulation concentrations of these heavy metals came from similar pollution sources [66], [67]. Results of Pearson's correlation coefficients and their significance levels ($P<0.01$) of correlation analysis were shown in Table 5. The concentrations of Cu, Zn, Pb, Ni and Cd showed strong positive relationship ($P<0.01$) with each other. This shows that Cu, Zn, Pb, Ni and Cd come from the same source. However, the concentration of Mn showed very weak correlations with the concentrations of the other metals, except Ni. This indicates that Mn have different sources than Cu, Zn, Pb and Cd. Han et al [68] also found the similar result about Mn by multivariate analysis.

Table 5: Correlations between heavy metal concentrations

	Cu	Zn	Pb	Ni	Mn	Cd
Cu	1.000					
Zn	0.919**	1.000				
Pb	0.870**	0.824**	1.000			

Ni	0.499**	0.524**	0.472**	1.000		
Mn	0.115	0.270	0238	0.515**	1.000	
Cd	0.906**	0.885**	0.972**	0.488**	0.285	1.000

Levels of significance: *P,0.05; **P,0.01. doi:10.1371/journal. pone.0039690.t005

Spatial Distribution of Heavy Metals in the Sediments of Jinzhou Bay

Geostatistics is increasingly used to model the spatial variability of contaminant concentrations and map them using generalized least-squares regression, known as kriging [69]–[71]. The probability map produced based on kriging interpolation and kriging standard deviation integrates information about the location of the pollutant source and transport process into the spatial mapping of contaminants [71], [72]. There are a lot of studies of the performance of the spatial interpolation methods, but the results are not clear-cut [73]. Some of them found that the kriging method performed better than inverse distance weighting (IDW) [74]; while others showed that kriging was no better than alternative methods [75]. For example, Kazemi and Hosseini [76] compared the ordinary kriging (OK) and other three spatial interpolation methods for estimating heavy metals in sediments of Caspian Sea, they found that the OK realization smoothed out spatial variability and extreme measured values between the range of observed minimum and maximum values for all of the contaminants.

The spatial distribution of metal concentrations is a useful tool to assess the possible sources of enrichment and to identify hot-spot area with high metal concentrations [48], [49]. Semivariogram calculation was conducted and experimental semivariogram of sediment heavy metal concentrations could be fitted with the Gaussian model for Cu, Zn, Pb, Cd, Ni and Mn. The theoretical variation function and experimental variation function exhibits a better fitting (Table 6). The values of R were significant at the 0.01 level by F test, which shows that the semivariogram models well reflect the spatial structural characteristics of sediemnt heavy metals.

Table 6: Parameters and F-test of fitted semivariogram models (Gaussian model) for heavy metals in sediment

	Nugget (C_o)	Sill (C_o+C)	C / (C_o+C)	Range	R^2	RSS	F test
Cu	5200	111500	0.953	0.0744	0.833	1.41E+09	24.94**
Zn	170000	4450000	0.962	0.0831	0.868	1.38E+12	32.88**
Pb	7900	106900	0.926	0.0883	0.837	7.23E+08	25 67**
Ni	10	7460	0.999	0.0277	0.513	3.81E+07	5.27**
Mn	7370	38310	0.808	0.0710	0.946	2.71 E+07	87.59**
Cd	220	5550	0.960	0.0935	0.921	8.49 E+05	58.29**
RI	890000	22880000	0.961	0.1004	0.927	1.12E+13	63.49**

**Significance at a = 0.01 level of F test. doi:10.1371/journal. pone.0039690.t006

The estimated maps of Cu, Zn, Pb, Cd, Ni and Mn clearly identified that the river, where the Huludao Zinc Smelter (HZS) is located at, is the most important source of heavy metals except Mn (Fig. 2). Among these metals, Cu, Zn, Pb and Cd showed a very similar spatial pattern, with contamination hotspots located at the estuary area, and their concentrations decreased sharply with the distance farther away from the estuary, indicating that they were from the same sources. The concentration of Ni also showed a similar pattern with the concentrations of Cu, Zn, Pb and Cd, but it changed not as sharply as the latter one. However, the concentration of Mn showed a completely different pattern with the others, indicating the industrial activities are not the source of Mn in the Jinzhou bay, and it may be related to geological factor. The mineral constituents of Jinzhou bay consist mainly of hornblende, epidote and magnetite. The percentage of hornblende, which is rich in Mn element, varied from 24.78% up to 64.70% [77]. This confirms that the concentration of Mn comes from the geological sources. The feature of point sources lies in that the inputs of heavy metals occur over a finite period of time and may have been effectively retained in the sediments near the sources, rather than re-suspended and distributed uniformly throughout the region [66]. Distinct from point sources, metals from non-point sources are more uniformly distributed throughout the area [50]. The results of this study closely correspond to this association.

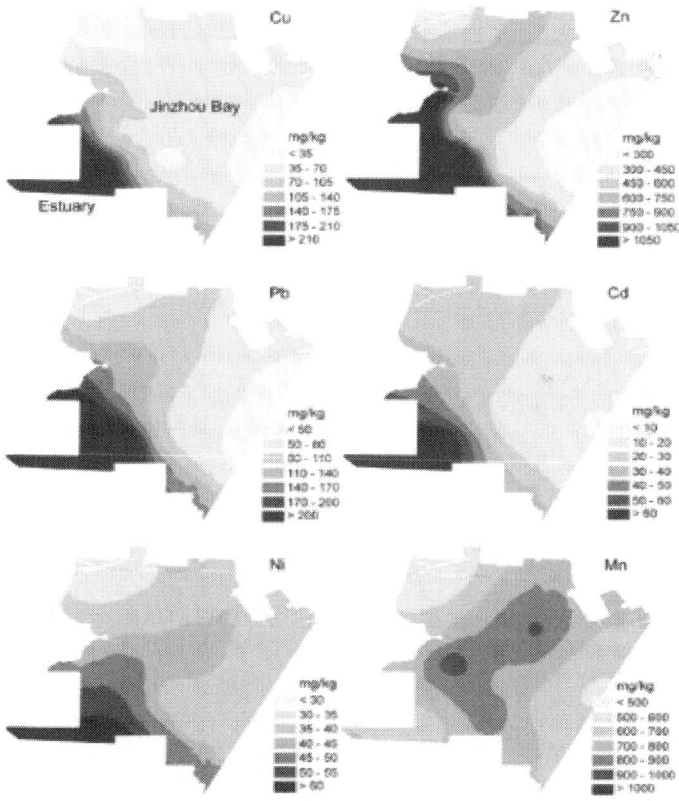

Figure 2: Estimated concentration maps for Cu, Zn, Pb, Cd, Ni and Mn (mg/kg).

The Spatial pattern of heavy metal in Jinzhou bay also provided a refinement and reconfirmation of the results in the statistical analysis, in which strong associations were found among Cu, Zn, Pb, Cd and Ni and very weak relations were found between Mn and the other heavy metals except Ni.

Assessment of Potential Ecological Risk, RI

Almost all the RI values of sampling sites were higher than 600 except the two samples from Wuli river, indicating that the sediments in the rivers of Huludao city and their estuary and Jinzhou bay exhibited very high ecological risk of heavy metals (Fig. 3, Table 7). The RI of sediments in the estuary was as high as 34.6 times of the line value for very high ecological risk level, suggesting that the sediments in the estuary were extremely polluted by heavy metals because of industrial discharge. Cd showed the highest potential ecological risk in the heavy metals, which contributed more than 95% of RI in the sampled sediments.

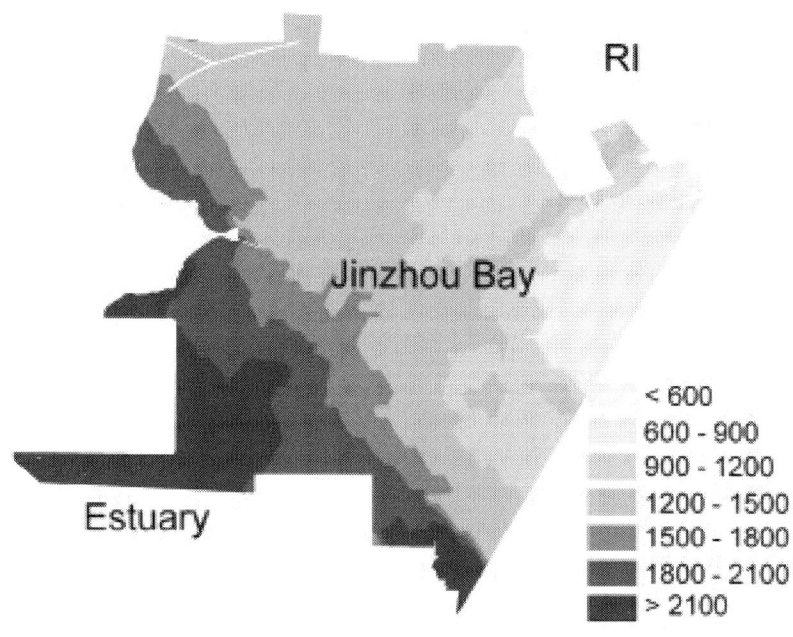

Figure 3: The spatial distribution pattern of RI of sediments in Jinzhou Bay.

Table 7: The heavy metal potential ecological risk indexes in sediments

Sediments	Er'						Pollution degree
	Cu	Zn	Pb	Ni	Cd	RI	
Lianshan River	16.64	4.23	9.36	31.89	3190.804	3252.96	very high
Wuli River	7.23	1.71	4.80	17.50	667.43	698.69	very high
Estuary	162.50	41.10	135.54	34.54	20414.20	20787.87	very high
linzhou Bay	10.59	4.60	10.33	24.15	1608.68	1658.34	very high

doi:10.1371/journal.pone.0039690.t007

Temporal Distribution of Heavy Metals in the Sediments of Jinzhou Bay

Sediment cores can be used to study the pollution history of aquatic ecosystem [44], [48]. Vertical distribution (0–36 cm) of heavy metals in Jinzhou Bay indicate that the concentration of Cu, Zn, Pb and Cd show similar vertical patterns (Fig. 4). The values of Cu, Zn, Pb and Cd increased sharply from the surface to its highest concentration at the depth of 8 cm and then decreased rapidly at a depth of 20 cm. The values of Cu, Zn, Pb and Cd varied slightly from a depth of 21–36 cm. The concentration of Ni in the sediment core decreased gradually from the surface with small fluctuations while the Mn remained relatively consent throughout the core.

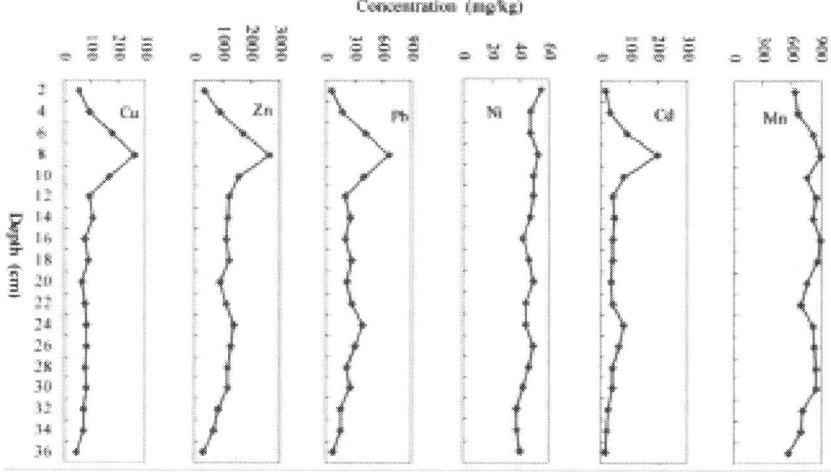

Figure 4: Vertical profiles of heavy metals for sediment core of Jinzhou bay, NE coast of China.

According to the sedimentation rate of about 1.0 cm/yr [78], the bottom of the sediment core (at the depth of 36 cm) was polluted about in 1973, 36 years after the set up of Huludao Zinc Smelter (HZS) in 1937. The heavy metals concentrations were already greatly higher than the national guideline values of marine sediment quality of China (GB 18668-2002), especially the concentration of Cd as high as 31 times of the latter. Related to annual Zn yields of HZS between 1973 and 2010 with the concentrations of Cu, Zn, Pb and Cd in sediment core, they showed similar temporal patterns, especially around the year 2000, the peak of Zn yield was followed the highest heavy metal concentrations at the depth of 8 cm of sediment core. This is conclusive evidence that Zn smelting operation was the dominant pollution source of aquatic environment in Jinzhou Bay. With the disposal and reuse of heavy metal wastewater from HZS around 2000, although the Zn yields increased year by year, the sediment pollution was alleviated gradually since 2000. The concentration of Ni in the sediment core decreased gradually from the surface with small fluctuations, the value varied from 54.36 mg/kg at the surface to 37.09 mg/kg at the depth of 32 cm. This also clearly indicated that the content of Ni in the sediment was come from the industrial discharges. Vertical profile of Mn shows some fluctuations in its concentration, and no obvious correlation with the depth of sediment core was observed. This indicates that the concentration of Mn is not under the control of human factors, as the result showed by the spatial pattern of Mn.

CONCLUSIONS

This study investigated the concentrations of heavy metals in the sediments from urban-stream, estuary and Jinzhou bay of the coastal industrial city, NE China. The results showed the impact of anthropogenic agents on abundances of heavy metals in sediments. The sediments are found to be extremely contaminated due to many years of random dumping of hazardous waste and free discharge of effluents by industries like Huludao Zinc Smelter (the largest zinc smelting plant in Asia), the Jinxi oil refinery and Jinhua chemical engineering, Huludao's massive shipyard and several arms factories. The potential ecological risk of sediments in lower river reaches, estuary and Jinzhou bay is at very high level, and Cd contributed more than 95% of RI in the sampled sediments. The estuary is the most polluted area, and its RI value was as high as 34.6 times of the line value for very high ecological risk level. The closer the distance to the estuary is, the higher RI values of sediments in Jinzhou bay are. The results of this research updated the information for effective environmental management in the industrial region. This study clearly highlights the urgent need to make great efforts to control the industrial discharges in the coastal area, and the immediate measures should be carried out to minimize the contaminations, and to prevent future pollution problems.

ACKNOWLEDGMENTS

Many thanks to the three anonymous reviews and the Academic Editor whose pertinent comments have greatly improved the quality of this paper and to Dr. V. Achal and Dr. X. Pan for lingual edit.

AUTHOR CONTRIBUTIONS

Conceived and designed the experiments: XL XH. Performed the experiments: XL LL BG. Analyzed the data: XL YW GL. Contributed reagents/materials/analysis tools: YW XC. Wrote the paper: XL XC XY.

REFERENCES

1. McKinley AC, Dafforn KA, Taylor MD, Johnston EL (2011) High Levels of Sediment Contamination Have Little Influence on Estuarine Beach Fish Communities. PLoS ONE 6(10): e26353. doi:10.1371/journal.pone.0026353.

2. Morton B, Blackmore G (2001) South China Sea. Marine Pollution Bulletin 42: 1236–1263.

3. Jha SK, Chavan SB, Pandit GG, Sadasivan S (2003) Geochronology of Pb and Hg pollution in a coastal marine environment using global fallout [137]Cs. Journal of Environmental Radioactivity 69: 145–157.

4. Muniz P, Danula E, Yannicelli B, Garcia-Alonso J, Medina G, et al. (2004) Assessment of contamination by heavy metals and petroleum hydrocarbons in sediments of Montevideo Harbour (Uruguay). Environment International 29: 1019–1028.

5. Xia P, Meng X, Yin P, Cao Z, Wang X (2011) Eighty-year sedimentary record of heavy metal inputs in the intertidal sediments from the Nanliu River estuary, Beibu Gulf of South China Sea. Environmental pollution 159: 92–99.

6. Fernandes C, Fontainhas-Fernandes A, Peixoto F, Salgado MA (2007) Bioaccumulation of heavy metals in Liza saliens from the Esomriz-Paramos coastal lagoon, Portugal. Ecotoxicology and Environmental Safety 66: 426–431.

7. Abdel-Baki AS, Dkhil MA, Al-Quraishy S (2011) Bioaccumulation of some heavy metals in tilapia fish relevant to their concentration in water and sediment of Wadi Hanifah, Saudi Arabia. African Journal of Biotechnology 10: 2541–2547.

8. Ridgway J, Breward N, Langston WJ, Lister R, Rees JG, et al. (2003) Distinguishing between natural and anthropogenic sources of metals entering the Irish Sea. Applied Geochemistry 18: 283–309.

9. Caeiro S, Costa MH, Fernandes F, Silveira N, Coimbra A, et al. (2005) Assessing heavy metal contamination in Sado Estuary sediment: An index analysis approach. Ecological Indicators 5: 151–169.

10. Sundaray SK, Nayak BB, Lin S, Bhatta D (2011) Geochemical speciation and risk assessment of heavy metals in the river estuarine sediments – a case study: Mahanadi basin, India. Journal of Hazardous Materials 186: 1837–1846.

11. Buccolieri A, Buccolieri G, Cardellicchio N, Dell'Atti A, Leo AD, et al. (2006) Heavy metals in marine sediments of Taranto Gulf (Ionian Sea, Southern Italy). Marine Chemistry 99: 227–235.

12. Voet E, Guinée JB, Udo de Haes H (2000) Heavy Metals: A Problem Solved? Methods and Models to Evaluate Policy Strategies for Heavy Metals. Kluwer, Dordrecht, the Netherlands.

13. Hjortenkrans D, Bergback B, Haggerud A (2006) New metal emission patterns in road traffic environments. Environmental Monitoring and Assessment 117: 85–98.

14. Govil PK, Sorlie JE, Murthy NN, Sujatha D, Reddy GLN, et al. (2008) Soil contamination of heavy metals in the Katedan Industrial Development Area, Hyderabad, India. Environmental Monitoring and Assessment 140: 313–323.

15. Wu SH, Zhou SL, Li XG (2011) Determining the anthropogenic contribution of heavy metal accumulations around a typical industrial town: Xushe, China. Journal of Geochemical Exploration 110: 92–97.

16. Snodgrass JW, Casey RE, Joseph D, Simon JA (2008) Microcosm investigations of stormwater pond sediment toxicity to embryonic and larval amphibians: variation in sensitivity among species. Environmental Pollution 154: 291–297.

17. Besser J, Brumbaugh W, Allert A, Poulton B, Schmitt C, et al. (2009) Ecological impacts of lead mining on Ozark streams: toxicity of sediment and pore water. Ecotoxicology and Environmental Safety 72: 516–526.

18. Ip CCM, Li XD, Zhang G, Wai OWH, Li YS (2007) Trace metal distribution in sediments of the Pearl River. Estuary and the surrounding coastal area, South China. Environmental Pollution 147: 311–323.

19. Suthar S, Arvind KN, Chabukdhara M, Gupta SK (2009) Assessment of metals in water and sediments of Hindon River, India: Impact of industrial and urban discharges. Journal of Hazardous Materials 178: 1088–1095.

20.   IGBP (1995) Global Change. In: Pernetta JC, Milliman JD, editors. Stockholm: ICSU.

21.   Cobelo-Garcia A, Prego R, Labandeira A (2004) Land inputs of trace metals, major elements, particulate organic carbon and suspended solids to an industrial coastal bay of the NE Atlantic. Water Research 38: 1753–1764.

22.   Birch GF, Taylor SE, Matthai C (2001) Small-scale spatial and temporal variance in the concentration of heavy metals in aquatic sediments: a review and some new concepts. Environmental Pollution 113: 357–372.

23.   Rubio B, Pye K, Rae JE, Rey D (2001) Sedimentological characteristics, heavy metal distribution and magnetic properties in subtidal sediments, Ria de Pontevedra, NW Spain. Sedimentology 48: 1277–1296.

24.   Sollitto D, Romic M, Castrignanò A, Romic D, Bakic H (2010) Assessing heavy metal contamination in soils of the Zagreb region (Northwest Croatia) using multivariate geostatistics. CATENA 80: 182–194.

25.   .Liu JG (2010) China's Road to Sustainability. Scinece 328: 50.

26.   .Pan K, Wang WX (2012) Trace metal contamination in estuarine and coastal environments in China. Science of the Total Environment 421–422: 3–16.

27.   NBSC (2010) National Bureau of Statistics of China. China Statistical Yearbook (2001–2009). Beijing.

28.   NBO National Bureau of Oceanography of China (2010) Bulletin of Marine Environmental Quality, 2008–2009.

29.   Yang ZF, Wang Y, Shen ZY, Niu JF, Tang ZW (2009) Distribution and speciation of heavy metals in sediments from the mainstream, tributaries, and lakes of the Yangtze River catchment of Wuhan, China. Journal of Hazardous Materials 166: 1186–1194.

30.   Fang TH, Li JY, Feng HM, Chen HY (2009) Distribution and contamination of trace metals in surface sediments of the East China Sea. Marine Environmental Research 68: 178–187.

31.   Müller B, Berg M, Yao ZP, Zhang XF, Wang D, et al. (2008) How polluted is the Yangtze river? Water quality downstream from the Three Gorges Dam. Science of the Total Environment 402: 232–247.

32. Chen C, Lu Y, Hong J, Ye M, Wang Y, Lu H (2010) Metal and metalloid contaminant availability in Yundang Lagoon sediments, Xiamen Bay, China, after 20 years continuous rehabilitation. Journal of Hazardous Materials 175: 1048–1055.

33. Zhang YF, Wang LJ, Huo CL, Guan DM (2008) Assessment on heavy metals pollution in surface sediments in Jinzhou Bay. Marine Environmental Science 2: 178–181.

34. Research and Markets website. Business monitor international. China Metals report. Accessed 2012 May 30.

35. Institute of Marine Environmental protection, State oceanic Administration (1984) Studies on the contamination and protection of Jinzhou Bay.

36. Lu CA, Zhang JF, Jiang HM, Yang JC, Zhang JT, et al. (2010) Assessment of soil contamination with Cd, Pb and Zn and source identification in the area around the Huludao Zinc Plan. Journal of Hazardous Materials 182: 743–748.

37. Zheng N, Wang QC, Liang ZZ, Zheng DM (2008) Characterization of heavy metal concentrations in the sediments of three freshwater rivers in Huludao City, Northeast China. Environmental Pollution 154: 135–142.

38. Wang J, Liu RH, Yu P, Tang AK, Xu LQ, et al. (2012) Study on the Pollution Characteristics of Heavy Metals in Seawater of Jinzhou Bay. Procedia Environmental Sciences 13: 507–1516.

39. Porstner U (1989) Lecture Notes in Earth Sciences (Contaminated Sediments). Springer Verlag, Berlin, 107–109.

40. Håkanson L (1980) An ecological risk index for aquatic pollution control: A sedimentological approach. Water Research 14: 975–1001.

41. Selvaraj K, Mohan Ram V, Piotr S (2004) Evaluation of Metal Contamination in Coastal Sediments of the Bay of Bengal, India: Geochemical and Statistical Approaches. Marine Pollution Bulletin 49: 174–185.

42. Verca P, Dolence T (2005) Geochemical Estimation of Copper Contamination in the Healing Mud from Makirina Bay, Central Adriatic. Environment International 31: 53–61.

43. Yi YJ, Yang ZF, Zhang SH (2011) Ecological risk assessment of heavy metals in sediment and human health risk assessment of

heavy metals in fishes in the middle and lower reaches of the Yangtze River basin. Environmental Pollution 159: 2575–2585.

44. Harikumar PS, Nasir UP (2010) Ecotoxicological impact assessment of heavy metals in core sediments of a tropical estuary. Ecotoxicology and Environmental Safety 73: 1742–1747.

45. Huang YL, Zhu WB, Le MH, Lu XX (2011) Temporal and spatial variations of heavy metals in urban riverine sediment: An example of Shenzhen River, Pearl River Delta, China. Quaternary International. doi:10.1016/j.quaint.2011.05.026.

46. Uluturhan E, Kontas A, Can E (2011) Sediment concentrations of heavy metals in the Homa Lagoon (Eastern Aegean Sea): Assessment of contamination and ecological risks. Marine Pollution Bulletin 62: 1989–1997.

47. Rosales-Hoz L, Cundy AB, Bahena-Manjarrez JL (2003) Heavy metals in sediment cores from a tropical estuary affected by anthropogenic discharges: Coatzacoalcos estuary. Estuarine, Coastal and Shelf Science 58: 117–126.

48. Karbassi AR, Nabi-Bidhendi GHR, Bayati I (2005) Environmental geochemistry of heavy metals in a sediment core off Bushehr, Persian Gulf. Iranian Journal of Environmental Health Science & Engineering 2: 255–260.

49. Viguri JR, Irabien MJ, Yusta I, Soto J, Gomez J, et al. (2007) Physico-chemical and toxicological characterization of the historic estuarine sediments. A multidisciplinary approach. Environment International 33: 436–444.

50. Shine JP, Ika RV, Ford TE (1995) Multivariate statistical examination of spatial and temporal patterns of heavy-metal contamination in New-Bedford Harbor marine-sediments. Environmental Science and Technology 29: 1781–1788.

51. White HK, Xu L, Lima ANL, Egliton TI, Reddy CM (2005) Abundance, composition and vertical transport of PAHs in marsh sediments. Environmental Science and Technology 39: 8273–8280.

52. Berthelsen BO, Steinnes E, Solberg W (1995) Heavy metal concentrations in plants in relation to atmospheric heavy metal deposition. Journal of Environmental Quality 24: 1018–1026.

53. Gray CW, McLaren RG, Roberts AHC (2003) Atmospheric accessions of heavy metals to some New Zealand pastoral soils. Science of the Total Environment 305: 105–115.

54. Li LL, Yi YL, Wang YS, Zhang DG (2006) Spatial distribution of soil heavy metals and pollution evaluation in Huludao City. Chinese Journal of Soil Science 37: 495–499.

55. Feng H, Cochran JK, Lwiza H, Brownawell BJ, Hirschberg DJ (1998) Distribution of heavy metal and PCB contaminants in the sediments of an urban estuary: The Hudson River. Marine Environmental Research. 45: 69–88.

56. Pekey H (2006) Heavy metal pollution assessment in sediments of the Izmit Bay, Turkey. Environmental Monitoring and Assessment 123: 219–231.

57. Raju K, Vijayaraghavan K, Seshachalam S, Muthumanickam J (2011) Impact of anthropogenic input on physicochemical parameters and trace metals in marine surface sediments of Bay of Bengal off Chennai, India. Environmental Monitoring and Assessment 177: 95–114.

58. Valdés J, Vargas G, Sifeddine A, Ortlieb L, Guiñez M (2005) Distribution and enrichment evaluation of heavy metals in Mejillones Bay Northern Chile: Geochemical and statistical approach. Marine Pollution Bulletin 50: 1558–1568.

59. Gomes F, Godoy J, Godoy M, Carvalho Z, Lopes R, et al. (2009) Metal concentrations, fluxes, inventories and chronologies in sediments from Sepetiba and Ribeira Bays: A comparative study. Marine Pollution Bulletin 59: 123–133.

60. Alba MD, Galindo-Riaño MD, Casanueva-Marenco MJ, García-Vargas M, Kosorc CM (2011) Assessment of the metal pollution, potential toxicity and speciation of sediment from Algeciras Bay (South of Spain) using chemometric tools. Journal of Hazardous Materials 190: 177–187.

61. Benoit G, Wang EX, Nieder WC, Levandowsky M, Breslin VT (1999) Sources and history of heavy metal contamination and sediment deposition in Tivoli South Bay, Hudson River, New York. Estuaries 22: 167–178.

62. Jonathan M P, Stephen-Pichaimani V, Srinivasalu S, RajeshwaraRao N, Mohan SP (2007) Enrichment of trace metals in surface

sediments from the northern part of Point Calimere, SE coast of India. Environmental Geology 55: 1811–1819.

63. Romano E, Ausili A, Zharova N, Magno MC, Pavoni B, et al. (2004) Marine sediment contamination of an industrial site at Port of Bagnoli, Gulf of Naples, Southern Italy. Marine Pollution Bulletin 49: 487–495.

64. Zhou F, Guo HC, Hao ZJ (2007) Spatial distribution of heavy metals in Hong Kong's marine sediments and their human impacts: a GIS-based chemometric approach. Marine Pollution Bulletin 54: 1372–84.

65. Al-Khashman OA, Shawabkeh RA (2006) Metals distribution in soils around the cement factory in southern Jordan. Environmental Pollution 140: 387–394.

66. Facchinelli A, Sacchi E, Mallen L (2001) Multivariate statistical and GIS-based approach to identify heavy metal sources in soils. Environmental Pollution 114: 313–324.

67. Manta DS, Angelone M, Bellanca A, Neri R, Sprovieri M (2002) Heavy metals in urban soils: a case study from the city of Palermo (Sicily), Italy. The Science of the Total Environment 300: 229–243.

68. Han Y, Du P, Cao J, Posmentier ES (2006) Multivariate analysis of heavy metal contamination in urban dusts of Xi'an, Central China. Science of the Total Environment 355: 176–186.

69. Carlon C, Critto A, Marcomini A, Nathanail P (2001) Risk based characterisation of contaminated industrial site using multivariate and geostatistical tools. Environmental Pollution 111: 417–427.

70. Romic M, Romic D (2003) Heavy metals distribution in agricultural topsoils in urban area. Environmental Pollution 43: 795–805.

71. McGrath D, Zhang CS, Carton OT (2004) Geostatistical analyses and hazard assessment on soil lead in Silvermines area, Ireland. Environmental Pollution 127: 239–248.

72. Saito H, Goovaerts P (2001) Accounting for source location and transport direction into geostatistical prediction of contaminants. Environmental Science & Technology 35: 4823–4829.

73. Xie YF, Chen TB, Lei M, Yang J, Guo QJ, et al. (2011) Spatial distribution of soil heavy metal pollution estimated by different

interpolation methods: Accuracy and uncertainty analysis. Chemosphere 82: 468–476.

74. Yasrebi J, Saffari M, Fathi H, Karimian N, Moazallahi M, et al. (2009) Evaluation and comparison of ordinary kriging and inverse distance weighting methods for prediction of spatial variability of some soil chemical parameters. Research Journal of Biological Sciences 4: 93–102.

75. Gotway CA, Ferguson RB, Hergert GW, Peterson TA (1996) Comparison of kriging and inverse-distance methods for mapping soil parameters. Soil Science Society of America Journal 60: 1237–1247.

76. Kazemi SM, Hosseini SM (2011) Comparison of spatial interpolation methods for estimating heavy metals in sediments of Caspian Sea. Expert Systems with Applications 38: 1632–1649.

77. Compiling Council of Chinese Embayment (1997) Chinese Embayment (Part 2). Beijing: China Ocean Press.

78. Ma JR, Shao MH (1994) Variation in heavy metal pollution of offshore sedimentary cores in Jinzhou Bay. China Environmental Science 14: 22–29.

Citations

CHAPTER 1

H. Sun, J. Li and X. Mao, "Heavy Metals' Spatial Distribution Characteristics in a Copper Mining Area of Zhejiang Province," Journal of Geographic Information System, Vol. 4 No. 1, 2012, pp. 46-54. doi: 10.4236/jgis.2012.41007.

CHAPTER 2

Baerbel Langmann, "Volcanic Ash versus Mineral Dust: Atmospheric Processing and Environmental and Climate Impacts," ISRN Atmospheric Sciences, vol. 2013, Article ID 245076, 17 pages, 2013. doi:10.1155/2013/245076.

CHAPTER 3

Iheoma M. Adekunle, Augustine O. O. Igbuku, Oke Oguns and Philip D. Shekwolo (2013). Emerging Trend in Natural Resource Utilization for Bioremediation of Oil — Based Drilling Wastes in Nigeria, Biodegradation - Engineering and Technology, Dr. Rolando Chamy (Ed.), ISBN: 978-953-51-1153-5, InTech, DOI: 10.5772/56526.

CHAPTER 4

V. Kirtskhalia, "Speed of Sound in Atmosphere of the Earth," Open Journal of Acoustics, Vol. 2 No. 2, 2012, pp. 80-85. doi: 10.4236/oja.2012.22009.

CHAPTER 5

Maxwell, A. and Strager, M. (2013) Assessing landform alterations induced by mountaintop mining. Natural Science, 5, 229-237. doi: 10.4236/ns.2013.52A034.

CHAPTER 6

Z. Li, C. Chen and J. Liu, "The Cyclic Behavior of Mountain Gravity Waves Generated by Flow over Topography," International Journal of Geosciences, Vol. 4 No. 3, 2013, pp. 558-563. doi: 10.4236/ijg.2013.43051.

CHAPTER 7

Gholizadeh A, Bor vka L, Vašát R, Saberioon M, Klement A, et al. (2015) Estimation of Potentially Toxic Elements Contamination in Anthropogenic Soils on a Brown Coal Mining Dumpsite by Reflectance Spectroscopy: A Case Study. PLoS ONE 10(2): e0117457. doi:10.1371/journal.pone.0117457.

CHAPTER 8

Li X, Liu L, Wang Y, Luo G, Chen X, et al. (2012) Integrated Assessment of Heavy Metal Contamination in Sediments from a Coastal Industrial Basin, NE China. PLoS ONE 7(6): e39690. doi:10.1371/journal.pone.0039690.

Index